Innovative Solutions

Emerging markets and emerging design approaches both present unique challenges for designing products and services, such as design inclusivity, cultural differences, the impact of new technologies, such as AI, care for the planet and evolving consumer needs. This newly updated book moves beyond usability and delight and explores the ethical, cultural, and systemic dimensions of user experience (UX) design. By understanding the complexities of new design approaches and technologies and their application in emerging markets, designers can better meet the demands of diverse user bases and create impactful and value-based solutions. An essential resource for those seeking to understand and innovate within emerging markets using new design approaches, it features chapters from leading experts on topics such as lived experiences, human values, social justice, sustainability, and the critical questioning of design's purpose. Contributions from local designers and researchers across India, Malaysia, Africa, Europe, the United States, New Zealand, and beyond provide practical insights into UX design and research in underserved communities and innovative tools such as ecosystem mapping.

Rather than prescribing a single framework, this book presents a tapestry of diverse perspectives. Contributors explore participatory methods, indigenous knowledge systems, AI ethics, and cross-disciplinary integration. The authors envision UX as a force for equity, inclusion, and systemic transformation. The future of UX, the book argues, is not fixed—but plural, situated, and transformative.

This second edition of *Innovative Solutions: Advanced User Experience Design* is ideal for professionals, researchers, and students of UX, human factors, product and systems design, design engineering, and human–computer interaction.

Innovative Solutions
Advanced User Experience Design

Second Edition

Edited by
Apala Lahiri
Girish Prabhu
Eric Schaffer

CRC Press
Taylor & Francis Group
Boca Raton London New York

CRC Press is an imprint of the
Taylor & Francis Group, an **informa** business

Designed cover image: Mamata Volvoikar

First edition published 2010
by CRC Press
2385 NW Executive Center Drive, Suite 320, Boca Raton FL 33431

and by CRC Press
4 Park Square, Milton Park, Abingdon, Oxon, OX14 4RN

CRC Press is an imprint of Taylor & Francis Group, LLC

ISBN: 978-1-032-90312-5 (hbk)
ISBN: 978-1-032-90395-8 (pbk)
ISBN: 978-1-003-55777-7 (ebk)

DOI: 10.1201/9781003557777

Typeset in Times
by SPi Technologies India Pvt Ltd (Straive)

Contents

SECTION 1   Design Frameworks and Practice: New Questions and Approaches

SECTION 2   Design and Research: Cultural Perspectives

SECTION 3   Enriching Design Practice through Multidisciplinary Collaboration and Integration

SECTION 4   Making Design Truly Human and Planet Centred: Learning from Practitioners

Preface

Over the last two decades, UX design has evolved from the periphery of software development to become a defining force in how people experience both digital and physical systems. With this rise comes not only unprecedented opportunities but also profound and often unexamined responsibilities. As designers increasingly shape access to health, education, governance, and community, the frameworks that guide UX practice demand renewed scrutiny.

This book was born from such scrutiny.

One of its seeds lies in Dr Apala Lahiri's doctoral research, which takes a critical lens to dominant UX paradigms. Her concept of **solidarity capital** reimagines the designer–user relationship—not as transactional, but as transformative. It calls on designers to engage with users not as abstract personas, but as individuals embedded in complex systems of care, justice, and meaning. Her work challenges us to view design as a moral and political act—particularly in high-stakes domains like healthcare and governance, where experience is inseparable from lived realities and institutional power.

Inspired by this foundational critique, Dr Eric Schaffer, Dr Apala Lahiri, and I invited contributors from across the world to reflect on where UX design is headed—and more importantly, where it *needs* to go. What emerged is not a singular manifesto, but a textured, at times conflicting, tapestry of perspectives. We view this diversity not as a limitation, but as a strength. We believe the future of UX is **plural**—not a fixed destination, but a field constantly reshaping itself in dialogue with shifting contexts, urgent challenges, and diverse human needs.

The book is organized into four thematic sections, each representing a different vector of inquiry.

SECTION 1:   DESIGN FRAMEWORKS AND PRACTICE: NEW QUESTIONS AND APPROACHES

This section interrogates the foundations of UX practice and theory. The contributors move beyond the limits of usability and delight, asking urgent questions about what design serves, whose voices it centers, and what futures it makes possible—or impossible.

- Dr Apala Lahiri reframes UX through **solidarity capital**, challenging the dominant logic of efficiency with a call to design for dignity, care, and co-agency. Her chapter emphasizes the ethical void in much of conventional UX, urging us to reclaim UX as a practice of social responsibility.
- Dr Girish Prabhu's **Culturally Appropriate Responsible Design (CARD)** framework critiques the narrow focus on individual users and proposes systems that embed cultural context, sustainability, and community well-being. Their work aligns UX with systemic design and intergenerational responsibility.

- Dr Eric Schaffer looks forward to the ethical and existential dilemmas of AI in UX. He warns of persuasive systems that exploit human vulnerabilities and advocates for a UX practice focused on experience, identity, and mental health in a post-work, algorithm-driven world. His chapter calls for a reorientation—from influence to integrity.
- Arvind Lodaya reclaims *Jugaad*—often dismissed as "frugal improvisation"—as a radical act of decolonial design. He argues for protecting jugaad as an indigenous form of innovation and calls for design justice and technological sovereignty in the face of institutional exclusion.
- Rosa Zubizarreta introduces **Dynamic Facilitation**, a methodology grounded in empathic listening, equitable participation, and psychological safety. Her approach shifts the power dynamics of UX research from "power over" to "power with," cultivating authentic insight through facilitative co-creation.

Together, these chapters reject "UX-as-usual." They position design not as a neutral process of interface-making, but as a political, ethical, and cultural act. They expand our vocabulary, challenge our frameworks, and illuminate what UX could become when it rises to meet its full societal responsibility.

SECTION 2: DESIGN AND RESEARCH: CULTURAL PERSPECTIVES

If Section 1 critiques prevailing UX logics, Section 2 brings cultural and geographic specificity to that critique. It explores how speculative, participatory, and situated methods can help build inclusive futures for historically marginalized populations.

- Abiodun Ogunyemi, Nicola Bidwell, Merja Bauters, Aderonke Sakpere, Nobert Jere, and Anicia Peters reposition speculative UX as a decolonial tool for Indigenous and African futures. Grounded in Ubuntu and African relational philosophies, their chapter offers a synthesis of Schwartz's values framework and community-centered ethics to shape just and sustainable technology imaginaries.
- Daniel Cabezas-López, José Abdelnour-Nocera, and Enric Mor provide cross-case insights from Nepal and East Timor, where **Participatory Design for Development (PD4D)** and **HCI4D** support inclusive, trust-building m-learning solutions. Their work critiques colonial models of technology transfer and champions co-created, contextually relevant educational design.
- G.H. Galal-Edeen, Amit Gudadhe, and colleagues present a mHealth UX case from Egypt's Base of the Pyramid populations. Their chapter emphasizes visual communication, inclusive tools for low-literacy users, and alignment with national healthcare strategy—demonstrating how UX can interface with policy and infrastructure.
- Sumita Sharma introduces **critical AI literacy** through participatory workshops with children in India, Japan, and Finland. Children's imagined AI

systems reflect values of fairness, diversity, and care—reminding us that UX must include youth not just as users, but as ethical co-designers of digital futures.

Together, these chapters expand UX beyond usability and efficiency—toward justice, imagination, and deep inclusion.

SECTION 3:   ENRICHING DESIGN PRACTICE THROUGH MULTIDISCIPLINARY COLLABORATION AND INTEGRATION

This section argues that UX can no longer afford to operate in disciplinary silos. The most impactful innovations emerge at the intersections of design, psychology, data ethics, public health, and education.

- Rama Vennelakanti and Joshua Ekandem advocate for emotional intelligence and inclusive leadership in AI-augmented UX. They call for a recalibration of design priorities—placing dignity, fairness, and empathy on par with speed and automation.
- Dr Owen Schaffer applies game design theory and positive psychology to build empirical frameworks for enjoyable user experiences. His work on **flow**, challenge-skill balance, and feedback mechanisms has implications not only for entertainment but also for learning and therapeutic UX applications.
- George Mathew examines the changing nature of designer-user relations in AI-driven systems. His chapter introduces concepts of co-agency and shared cognition, urging UX practitioners to address explainability, trust, and bias in intelligent interfaces. UX, he argues, must evolve from surface interaction to systemic co-creation.

Together, these chapters expand UX to include affective intelligence, behavioral science, and ethical foresight—offering designers deeper tools to engage with the emerging landscape of human–machine interaction.

SECTION 4:   MAKING DESIGN TRULY HUMAN AND PLANET CENTERED: LEARNING FROM PRACTITIONERS

s final section brings together deeply reflective voices from seasoned practitio- and scholars who have shaped UX and design thinking across sectors and geog- ᵉs. The authors offer rich insights into how design can become more relational, ic, and ethically anchored.

ab Jain urges designers to embrace ambiguity and cultivate imagination haping plural futures. Her work at Superflux shows how speculative ᵉn and storytelling can provoke new ways of being in the face of the ᵉ crisis and algorithmic governance.

- Ghada Refaat El Said centers UX in the Arabic-speaking world, advocating for linguistically and culturally embedded design. Her work spans mHealth, e-learning, and e-commerce—underscoring participatory design as a path toward ethical and localized innovation.
- Elisa del Galdo draws on decades of global UX research to stress the importance of trust, dignity, and inclusion. Her advocacy for a "slow design" approach highlights the necessity of contextual sensitivity and human difference.
- Joan Vinyets integrates anthropology and design thinking to reimagine institutional systems, particularly in healthcare. His work shows how deep listening and ethnographic insight can transform patient care, especially in sensitive settings like pediatric hospitals.
- Sudhindra V. brings a strategic lens to UX in large-scale digital systems. His emphasis on storytelling, foresight, and humility reflects a design leadership model grounded in complexity and transformation—not just usability or output.
- Alvin Yeo shares a remarkable journey of working with Indigenous and rural communities in Malaysia and New Zealand through participatory action research. Yeo emphasizes long-term, trust-based relationships as the foundation for impactful design. His work illustrates how participatory and culturally respectful ICT design can support digital inclusion, language preservation, and economic participation—while upholding data sovereignty and community control.

Together, these chapters anchor the book in real-world challenges. They advocate for a UX practice that is politically conscious, deeply participatory, and future-facing—one that honors lived experience and dares to imagine otherwise.

The contributors represent a broad spectrum of geographies—India, the United States, New Zealand, Australia, Egypt, Namibia, Finland, Spain, and the United Kingdom—and come from academia, industry, research labs, and consultancies. What unites them is a shared commitment to using UX as a lens not only for usability or satisfaction, but for **social impact and systemic transformation**.

CONCLUSION: A PLURAL, SITUATED, AND TRANSFORMATIVE UX

This book is not a blueprint. It does not offer a singular UX framework, nor does it promise easy answers. Instead, it presents a polyphonic map—charting where the field is headed, and more importantly, where it ought to go if it is to serve society in equitable, meaningful, and sustainable ways. In doing so, it welcomes divergence, embraces friction, and holds space for emergence.

This volume both expands and challenges the field of UX design. It expands by globalizing the discourse in substantive ways—bringing into focus underrepresented regions such as Egypt, Malaysia, Africa, India, and New Zealand. These are not presented as case studies in support of Western frameworks, but as vital sources of indigenous, situated knowledge and practice. It introduces and extends value-based frameworks such as CARD, solidarity capital, and critical AI literacy—adding to contemporary discussions on ethical and responsible AI and UX, yet grounding them in lived, field-tested experience. It also exemplifies the power of trans-disciplinary integration—drawing from anthropology, game design, public health, education, and policy to suggest what a richer, more responsive UX practice might look like. At the same time, the book challenges dominant paradigms. It critiques the instrumentalist, de-contextualized, and market-driven frameworks that have come to define mainstream UX—those often championed by UX consulting companies and corporate UX teams. It calls out the epistemic violence implicit in the uncritical export of universal heuristics, tools, and metrics to non-Western or underserved contexts—contexts where such assumptions erase difference and reinforce inequity.

The chapters within this volume often resonate with one another—but also surface productive tensions. Many contributors—explicitly or implicitly—advocate for systemic thinking, ethical reflection, human dignity, and participatory or decolonial approaches. They view design as a facilitator of dialogue and systemic inquiry, privileging ambiguity over certainty, multiplicity over monoliths, and engagement over control. Emotion, flow, and meaning are treated not as outcomes to optimize, but as dynamic states to honor and co-create. Almost all the chapters affirm a central tenet: that meaningful UX emerges through deep, long-term engagement with local context and community knowledge. At the same time, this diversity of contributions reveals critical tensions:

- between speculative and pragmatic approaches,
- between design as critique and design as intervention,
- between empirical validation and value-based meaning-making,
- between AI-enhanced design and human-centered judgment.

These tensions are not contradictions—they are the book's strength. They reflect the complexity of working across geographies, disciplines, and lived experiences. They remind us that design must hold multiple truths: that it can reform broken systems and imagine new ones; that it must scale and remain contextually grounded; that it can be rigorous and relational. In this way, the book joins a growing global movement—from Design Justice, Feminist HCI, and Postcolonial Computing, to Responsible Innovation, Indigenous Design, and Critical AI Studies—that calls for a more reflective, plural, and justice-oriented UX practice.

As editors, we believe that the future of UX does not lie in universal heuristics or generic toolkits. It lies in cultivating the capabilities to listen deeply, co-create wisely, and design courageously—with humility toward context, openness to critique, and an enduring commitment to both human and planetary well-being.

We hope this book invites you to take part in that future.

On behalf of all the editors and contributors, thank you for joining us in this inquiry.

Dr Girish Prabhu, Ph.D
Co-editor
Co-Founder, NISHThA.network

Contributors

José Abdelnour-Nocera
University of West London
London, UK

Merja Bauters
Tallinn University, School of Digital Technologies
Tallinn, Harju County, Estonia

Nicola J Bidwell
Charles Darwin University
Darwin, Northern Territory, Australia
Rhodes University
Makhanda, Eastern Cape, South Africa

Luc Bolier
The Hague University of Applied Science
The Hague, The Netherlands

Daniel Cabezas-López
Universitat Oberta de Catalunya (UOC)
Barcelona, Spain

Po-Ying Chao
Packaging101
Taichung, Taiwan

Elisa del Galdo
Del Galdo Consulting
London, UK

Joshua Ekandem
GRWNDED
Santa Rosa, CA, USA

Galal H. Galal-Edeen
American University in Cairo & Cairo University
Cairo, Egypt

Amit Gudadhe
Hilti
Paris, France

Anab Jain
Superflux Studio
London, UK

Nobert Jere
University of Fort Hare, Computational Sciences Department
Alice, South Africa

Apala Lahiri
HFI
Fairfield, USA

Arvind Lodaya
Bengaluru, India

George Mathew
LexX Technologies Pty. Ltd
Sydney, Australia

Enric Mor
Universitat Oberta de Catalunya (UOC)
Barcelona, Spain

Abiodun Ogunyemi
Tallinn University, School of Digital Technologies
Tallinn, Harju, Estonia

Anicia Peters
National Commission on Research, Science and Technology (NCRST)
Windhoek, Namibia

Girish Prabhu
NISHThA.network / DesignFold
Bengaluru, India

Ghada Refaat El Said
Future University
Cairo, Egypt

Joan Vinyets i Rejón
Hospital Sant Joan de Déu
Barcelona, Cataloia, Spain

Aderonke Sakpere
University of Ibadan, Faculty of Computing
Ibadan, Nigeria

Eric Schaffer
HFI
Fairfield, USA

Owen Schaffer
Elmherst University
Chicago, USA

Sumita Sharma
INTERACT Research Unit
University of Oulu
Oulu, Finland

Ahmed Sorour
BBITS
Dubai, UAE

Sarah Suib
hint-studio
Brussels, Belgium.

V. Sudhindra
TCS, Australia
Melbourne, Victoria, Australia

Rama Vennelakanti
Independent
Bengaluru, Karnataka, India

Alvin W. Yeo
Software Engineering Dept,
    School of Computing and
    Mathematical Sciences,
    University of Waikato
Hamilton, Waikato, New Zealand

Rosa Zubizarreta
Senior Fellow at the Research Institute
    for Sustainability,
Potsdam, Germany

About the Editors

Dr Apala Lahiri is Co-founder of Human Factors International APAC and Global Chief of Technical Staff. She has led several strategic cross-cultural and innovation initiatives for MNCs across the world. She is a well-known speaker at major international conferences. Her passion is to envision how user experience can be a positive change agent using digital channels. She is also fascinated by changes in user experience across time and space. Apala is an award-winning designer (International Audi Design Award). She co-edited the book Innovative Solutions: What Designers Need to Know for Today's Emerging Markets. She was invited to speak on User Experience Design and AI at the AI for Good conference, 2017, organized by the United Nations. Apala's doctoral research explores the creation of a new UX framework that considers sustainability and societal issues as integral parts of the design process. She recently co-founded UX4.me, an AI-enabled service platform for UX research.

Dr Girish Prabhu, Co-Founder of NISHThA. network is a design strategist, educator, and innovation leader with over three decades of experience spanning industry, research, and academia. He has held senior leadership roles at Kodak, HP Labs, Intel, HFI, and the Manipal-Srishti Institute of Art, Design and Technology, where he has been instrumental in advancing human-centered design, UX research, and responsible innovation. With a polymathic and systems-thinking approach, Girish draws from a wide range of domains—including healthcare, wellness, education, consumer products, and retail—to shape design that is both context-sensitive and future-oriented. His work is grounded in the conviction that design must go beyond aesthetics and usability to engage with deeper questions of equity, inclusion, and long-term sustainability. A mentor to generations of designers and researchers, he has played a key role in shaping design education and innovation ecosystems in India and internationally. He is currently developing interdisciplinary leadership models through the Institute for Systemic & Polymathic Leadership, and continues to build frameworks for culturally grounded, ethically responsible design. As co-editor of this volume, he brings critical insight into the transformative potential of UX design to address complex societal challenges.

Dr Eric Schaffer is Founder and CEO of HFI, and Head of the HFI Laboratories. Eric has been working in the User Experience engineering field since 1977. He holds a doctorate from Steven's Institute in applied psychology, specializing in human performance, and he is a Board-Certified Professional Ergonomist. He worked at Mauro Associates, AT&T Long Lines, and Bell Labs, and then he founded Human Factors International, Inc., starting in 1981. While leading HFI, he has focused on building UX capabilities in organizations and delivering sustained UX teams for complex environments. Eric was the primary driver in developing the HFI courses, certifications, and toolsets. He recently co-founded UX4.me, an AI-enabled service platform for UX research. Eric has been a driver in the spread of UX engineering worldwide and helped pioneer remote and offshore operations. Eric lives in Fairfield, Iowa, and Pondicherry, India.

Section 1

Design Frameworks and Practice

New Questions and Approaches

1 Design of Everyday Solidarity

Apala Lahiri

INTRODUCTION

User experience (UX) design of digital spaces has become a highly sought-after sub-field in the last decade, growing rapidly worldwide. Today, the field of UX attracts many designers and non-designers, such as Infotech workers. Thus, the impact of UX design needs to be explored. My doctoral thesis, submitted to Srishti Manipal Academy of Higher Education in 2024, investigated and analyzed the major gaps created by dominant UX design frameworks and the approaches taken by alternative frameworks to address specific issues arising from the use of dominant UX design frameworks. Drawing on the work of various critical theorists (Bourdieu, Arendt, and Foucault) and design approaches, such as Equity-Centered Community Design (ECCD), Participatory Design (PD), and Design Justice, the concept of a UX professional's solidarity capital has emerged as a crucial construct to help address some of the identified gaps. Solidarity capital would facilitate a transformation in the relationship between the designer and the citizen (aka "user") from a fleeting, transactional dynamic between expert and subject to one in which the co-creators collaborate in collective spaces, making citizens an integral part of the human-centered design process.

GAPS IN UX DESIGN

The journey of user experience (UX) design has primarily involved close collaboration with technology, all the way from ergonomics to UX, and has been heavily influenced by psychology. Consequently, scientific methods dominated the field, marginalizing lived experiences unless they could be mathematically analyzed. This emphasis on precision distorted the view of experience. "The instruments of science were helpless in the realm of qualities. The qualitative was reduced to the subjective: the subjective was dismissed as unreal, and the unseen and unmeasurable non-existent" (Mumford 1923).

Based on a literature review of dominant and alternative UX design approaches and frameworks, three major gaps of current dominant UX practice emerged:

1. UX and the Banality of Evil
2. UX and the Objectification of Humanity
3. UX and the Myopia of Vision

DOI: 10.1201/9781003557777-2

These gaps highlight critical ethical and conceptual challenges that call for a rethinking of how UX professionals engage with users and society.

UX AND THE BANALITY OF EVIL

In "Eichmann in Jerusalem" (Arendt 1963), Hannah Arendt described what she called "the banality of evil." Arendt describes the horror of evil actions by ordinary people like Adolf Eichmann. Another interpretation suggests that she highlights how systems can condition normal individuals to play roles and follow guidelines, leading to ignorance of the broader implications. Essentially, a well-structured system allows individuals such as Eichmann to compartmentalize and overlook the moral consequences of their actions within the system.

Dominant UX frameworks yield considerably similar outcomes by treating designed experiences as a collection of tasks for the end user, with the user's efficiency in completing any task being the primary criterion for success and satisfaction. The emphasis on designing "experience" by deconstructing it into parts is evident in how the success of UX is assessed. Current UX frameworks dissect the "whole" and concentrate on the limited "experience" components based on tasks, goals, or needs, using appropriate metrics to evaluate these parts. However, none of them regard the larger ecosystem in which the user operates. In other words, defining successful UX in terms of the success of each component (or task/goal) rather than measuring the "whole" is the fundamental problem.

Current UX frameworks and metrics tend to reduce practitioners and "users" to being "banally evil" by dissecting the "whole" and constructing the experience in parts, achieving compartmentalized efficiency. Does this reductionist approach not diminish human potential, defining people by mere roles and functions measured by individual efficiency, speed, and satisfaction rather than considering societal impact and capability enhancement? For instance, a UX designer focusing on creating a pre-approved microloan experience on a banking site might aim to design a highly efficient virtual agent for users to secure their loans quickly.

What the designer may not realize, as they are looking at a part of the picture instead of the whole picture, is that the virtual agent also implicitly creates a channel for using nudges and pushes to provide, for instance, instant loan gratification to masses of people. These people can be taken too far forward, too quickly, in a possible virtual agent-initiated journey for instant micro loans to be unable to pause and reflect (and refuse the loan) and, thus, get lured into a possible continuous state of debt.

UX AND OBJECTIFICATION OF HUMANITY

UX is founded on the philosophy that "users" must be "sampled" from predefined categories within the relevant context or ecosystem. Marketers segment users based on how each group is likely to spend on products and services, using attributes such as demographic, psychographic, behavioral, and geographic factors. Sample size and selection criteria are established according to these segment definitions, leading to the design of UX research. Insights from this research inform optimal design solutions for different user categories. However, categorizing humanity into distinct, often hierarchical "user" segments is problematic, particularly when it primarily aligns with the needs of markets, governments, or organizations. Does this segmentation of "users" enforce a homogenized mental model for designers to aspire to? "A seemingly neat and simple way to reduce humanity to an average to be respected, or as an optimum toward which one must move. It measures in quantitative terms and hierarchizes in terms of abilities, the level, the nature of individuals. It introduces through this value giving measure, the constraint of a conformity that must be achieved" (Foucault 1984).

This segmentation also provides a clear view of who occupies the bottom of the socioeconomic pyramid. Consequently, the inclusion of participants in the research is typically based on their purchasing power and "the most common or most profitable group of users" (Baumer and Brubaker 2017). The Pareto Principle (Pareto 1971) is frequently cited as justification. Without market logic, persuading clients toward inclusivity is often ineffective. The key question is whether UX can be practiced differently. If true human-centeredness is central to UX, can we stop viewing the rich human spirit merely as segments of "Users" and instead design for humanity, prioritizing ethical needs and freedom over business-driven "opportunities" for segments? Can we continuously collaborate with those experiencing our designs to ensure their voices empower critical thinking and meet their desired experiences?

UX AND THE MYOPIA OF VISION

The field of UX and its processes continue to ignore the emergent "beam of darkness." (Agamben 2009, 45). As evidenced by the literature review (Chapanis 1979; Norman 2005; Kraus 2014), from the mid-1940s to today, the emphasis remains on a selective "slice" of the present, neglecting those slices that cast shadows on the future. Additionally, the dominant UX design frameworks focus not on the future impact of changes resulting from present design practices but on "opportunities" for the here and now.

REIMAGINING THE DESIGNER–CITIZEN RELATIONSHIP

The thesis emphasized a necessary shift in the designer–citizen (or "user") relationship, to enable significant transformations in dominant UX design frameworks and bridge existing gaps. Designers could continuously enrich their understanding of citizens' lived experiences, cultural values, and desired needs and capabilities by forming strong connections with citizens. This mutual learning would foster agency among citizen designers, empowering them within networks of solidarity. Consequently, designers could transcend mere problem-solving to become conscience keepers, questioning the purpose of design and its broader impacts on society and structural inequalities. Their collaborative efforts could yield ethical, innovative solutions that benefit businesses while partnering with citizen designers.

This expanded connection with citizens would encourage and enable designers to embrace broader responsibilities, promoting community strength and collective building. Such practices might create alternative spaces temporarily and forge new connections and social relations, potentially resulting in establishments such as Experience Design Coops, Collectives, and Commons.

This is especially important, given the often-discussed critique about design and design research, where an elite group of designers make decisions on behalf of institutional sponsors, leading the invisible "user" to live with these decisions (Rittel and

Webber 1973; Schon 1959; Papanek 1985). Several dissenting voices (Bannon and Ehn 2013; Dunne 1999; Manzini 2015; Papanek 1985) have raised concerns that "design, like mass media before it (e.g., the dream-factory of film, in Bloch's words), peddles in smooth surfaces and pleasant consumerism, while masking and ultimately furthering the interests of an undesirable and unsustainable socioeconomic order" (Tonkinwise 2013).

Some new approaches have been critiquing the dominant design frameworks and creating alternative design attitudes, which are discussed in the following subsections.

EQUITY-CENTERED COMMUNITY DESIGN (ECCD)

Sharing the philosophy of social impact design approaches, Equity-Centered Community Design (ECCD) "believes that a designer is anyone who has agency to make a decision, however small, that will impact a group of people or the environment" (Creative Reaction Lab n.d.).

ECCD, as described by Creative Reaction Lab,

> is a unique creative problem-solving process based on equity, humility-building, integrating history and healing practices, addressing power dynamics, and co-creating with the community. This design process focuses on a community's culture and needs so that they can gain tools to dismantle systemic oppression and create a future with equity for all. Creative Reaction Lab's goal is to share equity-centered community design to achieve sustained community health, economic opportunities, and social and cultural solidarity.

> (Creative Reaction Lab n.d.)

PARTICIPATORY DESIGN

The objective of the participatory design (PD) movement, which originated in Scandinavia in the 1960s and 1970s, was to involve various stakeholders—workers and managers—in the design process. This approach aimed to ensure that the final design solution would be informed by a deeper understanding of the stakeholders' work ecosystems (Bardzell 2010). "Participatory design differs from user centered design in the belief that all people have something to offer to the design process and that they can be both articulate and creative when given appropriate tools with which to express themselves" (Sanders 2002).

PD originated from a considerably strong political stance the researchers/ designers had in the 1980s (Bodker et al. 2000), when starting the co-operative design movement in Scandinavia. Political commitment to democracy in workplaces drove the researchers and designers involved at this time. They collaborated with local trade union members to ensure their participation in socio-technical change and influenced government policy to support democratic local processes. Worker participation in designing workplace technology was the main focus of early PD.

Early PD emphasized core principles ensuring that end users had a voice in system design and fostering mutual learning between designers and all stakeholders (Simonsen and Robertson 2013).

Design Justice

The use of community participation, along with creative and collaborative practices that acknowledge the local context in its historical significance and future direction, while considering the multiple levels of oppression faced by marginalized groups—especially in an intersectional context—forms the foundational philosophy of Design Justice.

Foregrounding Intent of Citizens

Reflecting on various approaches relevant to UX design, including ECCD, PD, and Design Justice, a key direction emerges: foregrounding the intent of all people impacted by technology and design, not merely dominant groups. This requires their participation in shaping outcomes. An alternative framework must ensure the involvement of all stakeholders—dominant and marginal—in the design and research process. Such continuous engagement can foster democratic decision-making, mutual learning, and prevent the imposition of a privileged minority's intent in system-building experiences.

BRIDGING DESIGN AND INFORMATION SYSTEMS: INSIGHTS FROM THE SOCIO-TECHNICAL PERSPECTIVE

After reviewing the literature on dominant and alternative UX frameworks, the thesis examined the potential for applicable frameworks and methodologies outside of design. The choice to investigate information systems arose from the significant impact of the socio-technical view on PD at its origin (Ehn and Kyng 1987). The socio-technical view is one of the four main definitions of Information Systems (Ghaffarian 2011). The term "socio-technical" was first introduced by researchers at the Tavistock Institute of Human Relations in the United Kingdom. This approach gained strong acceptance in Scandinavian countries, reflecting the principles of PD (Kuutti 1996). Key figures in the early development of PD, including Kristen Nygaard, Bo Dahlbom, Pelle Ehn, and Erik Stolterman, pioneered the Scandinavian socio-technical study of computing. The socio-technical approach to studying information systems and Information and Communication Technologies (ICT) is extremely different from other approaches in distinct ways, as specified in the next sections.

Inclusion of Context

This approach involves data collection over time rather than disregarding the temporal dimension and relying on a single snapshot data collection method, which is how approximately 90% of information systems research using other approaches operates (Orlikowski and Baroudi 1991). This approach leads to a holistic view of context by paying attention to all contextual elements and, hence, does not reduce "contextual insights into parsimonious explanations."

Unlike other approaches to information systems that prioritize rationality and lean toward economic and technological determinism to maximize utility as the primary goal, the socio-technical approach broadens the perspective and recognizes the subjectivity of various stakeholders. The definition of a stakeholder is also more inclusive. "Socio-technical research seeks to include other legitimate actors in the IS/ICT innovation scene: the operators, the customers, the citizens, the gendered individuals, the poor, the government and the like" (Avgerou, Ciborra and Land 2004).

The socio-technical perspective has led information systems research to shift toward critical research. Critical researchers often use critical social theories of Habermas, Foucault, and Bourdieu and concern themselves with issues such as "freedom, power, social control, and values with respect to the development, use, and impact of information technology" (Hirschheim and Klein 1994).

Myers and Klein have proposed a set of principles for critical research after reviewing major discourses and various approaches present in different forms within the field of critical research. The principles are (Myers and Klein 2011):

Element of Critique

- Principle of using core concepts from critical social theorists
- Principle of taking a value position
- Principle of revealing and challenging prevailing beliefs and social practices

Element of Transformation

- Principle of individual emancipation
- Principle of improvements in society
- Principle of improvements in social theories

They suggest that the principles they formulated build on the commonalities shared across multiple approaches within critical research in information systems. In addition to these principles, the following points reinforce the relevance of the critical research approach for any reimagining of mainstream UX design frameworks:

- Unlike positivist natural sciences, it gathers insights from how people make sense of their experiences.
- There is a willingness to critique dominant beliefs that seem universal and natural.
- The focus on transformation and envisioning how the world could be instead of merely interpreting it as it is (Stahl and Brooke 2008; Myers and Klein 2011).
- The appropriateness of critical research in contexts where the research method entails some form of intervention in the real or virtual world (Myers 1997; Thomas 1993). Additionally, a more holistic consideration of the ethical and social aspects of the interventions decided by "action researchers and design science researchers" is more likely to be enabled by critical research (Stahl and Brooke 2008; Myers and Klein 2011).
- The critical role of research in establishing a transformative perspective in the design and development of Information Systems (IS), a field also related to UX, cannot be overstated.

REVISITING CURRENT UX PRACTICE—THE COUNTER DISCOURSE

Core concepts from critical social theorists were employed to create a counter discourse aimed at bridging the gaps present in dominant UX frameworks.

Using Core Concepts From Pierre Bourdieu

Bourdieu's *Outline of a Theory of Practice* (1977), *Distinction* (1984), and *An Invitation to Reflexive Sociology* (Bourdieu and Wacquant 1992), and other publications (Bourdieu 1971, 1986) sought to shift away from "variable-centred hypothesis testing social science research towards his approach to social fields, capital and habitus" (Ignatow and Robinson 2017). Bourdieu examined social science research through a relational approach, viewing social life in context as part of a whole. His concept of habitus, for example, refers to the encoding of a specific cultural understanding and the internalization of the social order through patterns of thought and behavior and taste that an individual acquires while recognizing "the agent's practice, his or her capacity for invention and improvisation" (Macey 2000).

The spatial metaphor of "field" is also a distinctive part of Bourdieu's theory. A field, according to Bourdieu, orders social relations where economic, social, and cultural capital determine social position and relations. Bourdieu considers capital to be deeply connected with the concept of field. Capital, whether cultural, social, or symbolic, is, for Bourdieu, a reference to the "stocks of internalized ability and aptitude as well as externalized resources which are scarce and socially valued."

Similar to economic capital, each of these forms of capital can be exchanged and transformed. Bourdieu also views all forms of capital as "means to produce or reproduce inequality" (Bourdieu 1977, 1986; Calhoun et al. 1993; Harker et al. 1990).

Bourdieu's concept of social capital laid the groundwork for the idea of solidarity capital, which serves as the basis for the counter discourse. Social capital, according to Bourdieu (1986), represents a person's social relations. This network of social relations provides access to material and immaterial resources, resulting from exchanges of different forms of capital. For example, one can trade economic capital for cultural capital by obtaining a degree from a premium institution, which can help accumulate more economic capital and elevate social capital through connections with high-capital professors and classmates. Considering the positive and negative effects of social capital, the thesis argued that designers can counter the symbolic capital of business stakeholders in the corporate design ecosystem by enhancing a specific capital that business stakeholders rarely have access to. This capital is a blend of social capital and concepts from the social solidarity economy.

The social solidarity economy, according to Esteves (2019) and Laville (2016), focuses on the potentialization and increase of social capital as the "organizing factor of production," as well as its mobilization to create "civic capital" by directing it toward strengthening democratic functioning within organizations and across society. I define solidarity capital, based broadly on the construct of social capital and social solidarity economy, as the strength of the UX professional's solidarity network spaces that can be considered sites of "mobilizational citizenship" (Escoffier 2018). These design solidarity network spaces co-created by designers and citizens should focus on the making of productive communities which "create social power that constrains traditional business models and pressures them into some kind of adaptation with the expectations of (citizen) stakeholders" (Bauwens and Niaros 2017; Esteves 2019).

Solidarity capital for designers should counter the objectification of humanity that leads to excluding large populations from positive experiences. This capital, emphasizing solidarity networks that challenge traditional business models, bolsters a designer's symbolic capital by presenting a collective voice, to business stakeholders, of citizens who will influence how experiences are consumed. Since a collective represents many, not merely a few people and given the market's fascination with numbers (Mumford 1923), this collective that defines solidarity capital could be a vital leverage for shifting business mindsets. Therefore, designers with their innate sensitivity and the learning and practice of empathy are well-positioned to effect a shift in their power, along with the power of the agentic citizen, by building solidarity capital. This also entails transforming the relationship between designer and "user" from one of researcher and the researched to that of a solidarity network of designers and citizens working shoulder to shoulder in creating solidarity networks that function as "spaces of embodiment, coordination…as well as of their development through formal and informal practices of learning and knowledge production" (Rakopoulos 2015). These networks should, as they evolve, promote mobilizing communities through experience design manifestos for the technologies and experiences they do or do not desire.

Solidarity capital can symbolize the designer and her field's involvement in design discourse. By fostering citizen-centric dialogue from the project's outset, the designer can expand her solidarity capital and networks, encouraging stakeholders from sponsor organizations to harmonize their intent with that of citizens.

CHANGING SOCIETY BY CREATING CITIZEN DESIGNERS

The unanswered question was how designers could connect with citizens to understand their habitus and the Values, Needs, and Capabilities that shape it, along with any changes over time. Additionally, how would designers gauge citizens' awareness and opinions on crucial issues like inclusivity and sustainability to inform alternative solutions? Without addressing this, alternative frameworks would continue to prioritize the designer/business stakeholders' perspectives, based on a limited view of citizens' thoughts and feelings, as seen in dominant UX frameworks. Thus, it was essential to elaborate on how it was possible to create the designer's solidarity capital.

- How could designers and citizens engage in a two-way interaction that continuously enhanced their awareness and understanding of each other's fields, allowing them to jointly contribute to the objectives of a specific design initiative and define the outcomes?
- Could the collaboration of diverse individuals, as a collective, foster solidarity and a shared vision for a just and equitable world?
- Could citizen designers help to normalize inclusivity in the scope of the final designed experience?

This seemed impossible without the designer's ability to connect with citizens from all walks of life and continue evolving these connections over time, free from any specific market-driven agenda. From this solidarity-based collaboration, designers and citizens could develop a shared understanding of creating new knowledge and imaginaries needed for social change. Continuous interaction and collaboration could enhance our design practices, making them more aware and invested in addressing challenges such as inequality and global climate change.

Numerous inspiring examples from the social solidarity economy showcase citizen participation in various social solidarity collectives. This leads to the roles of consumers and citizens intermingling, creating new opportunities for collaboration between consumers and producers. The "gasistas" or members of the Italian Gruppi di Acquisto Solidale (GAS or Solidarity Purchase Groups), for example, maintain that they not only exercise ethical or critical consumption as individuals or as groups, but that they "co-produce" a commonwealth. The ongoing networking among gasistas facilitates social learning, building social capital, and multiplying the potential of individuals to act (Forno, Grasseni, and Signori 2015).

The concept of citizen designers is greatly inspired by the work done by the participatory and justice design movement. As John Rheinfrank (2002) pointed out almost two decades ago, "the designer is moving from being the detached expert to be a collaborator." This will imply that project goals will be defined by considering user views. Thus, the value of co-creation with users will gain momentum. "In the past we designed for users, today we design with users, tomorrow we will have design by users" (Rheinfrank 2002).

Taking inspiration from PD and design justice ecosystems, it became clear that for a real solidarity-based relationship between designers and citizens to be in place, the connection between designers and citizens and conversations between them cannot be mediated by others, such as clients or project sponsors. Designers must engage citizens across locations and segments for ongoing conversations, even if from a distance, in a practical, sustainable, and cost-effective way.

RETHINKING SOCIAL MEDIA AS RESEARCH SITES

BEYOND A DYSTOPIAN VISION OF WHATSAPP

Using social media to build solidarity networks is practical and cost-effective, as noted in literature on social media activism (Gil de Zúñiga, Ardèvol-Abreu, and Casero-Ripollés 2021; Emiliano Treré 2020). WhatsApp faces criticism for its role in spreading false information, hoaxes, hate speech, propaganda, and fake news in various political scenarios.

However, in the literature review, the authors describe the social media issues while presenting a more nuanced and ambivalent picture, illustrating how WhatsApp, for instance, has been integrated into the protest media ecologies and daily activities of movement organizations. WhatsApp (along with similar apps) has become a key ordinary digital ally, not only a tool that is "weaponized" during electoral times.

The result is that ordinary citizens are increasingly using mobile instant messaging apps such as WhatsApp for politically related activities. Compared to other "semi-public" online platforms, WhatsApp provides a more intimate and controlled environment in which users can almost simultaneously gather and share news, discuss politics, and mobilize others.

These platforms and apps are integral to people's daily lives and are embedded in almost everyone's familiar channels for connecting with others and sharing or receiving information. These are not elite channels and, therefore, are not restricted to a small privileged segment of the population. However, there is also the increasingly troubling aspect of these channels: data privacy, which remains a concern.

RESEARCH EXPERIMENT

An experiment using WhatsApp for research was conducted for a duration of two months, involving designers discussing banking and wealth with a group of citizens (who we called citizen designers), to understand their perspective. This research aimed for designers and citizens to become partners, fostering a solidarity-based relationship. The objective was to see if designers could expand their social capital through accessible social media and gain new insights into their role. Six designers from my UX practice and my co-editor, Dr. Girish Prabhu, volunteered for this research with me.

PARTICIPANT PROFILES

As there were no constraints imposed on which segments the citizen designers needed to belong to (as is the case in client-funded projects), broad criteria were used.

Adequate representation was sought based on:

- Gender
- Age (everyone from 16 + could be included)
- Life stage (student, employed/self-employed/unemployed, retired)
- Location—any place in India

ACTIVITIES

Eighteen citizen designers collaborated on the research, which lasted about two months and involved 19 questions. Despite the number of questions, the research took much longer because, in the spirit of a conversation that encouraged equal participation, some citizen designers needed more time and iterations to share their perspectives on specific questions. At times, some citizen designers became busy, leading to a break in the flow of conversation; other times, they sought clarification on the purpose of the question or the reason behind a sudden change in the direction of the discussion.

ANALYSIS

After 19 questions, I decided to pause and analyze the conversation's content up to that point. This analysis would guide how to share insights with citizen designers and continue the interaction. I used thematic analysis (Gibbs 2012; Saldaña 2014), as identifying emergent themes was more crucial than statistical insights. Clark, Braun, and Hayfield (2015) also emphasize that thematic analysis fits research questions focusing on people's perspectives, lived experiences, and the social construction of a topic.

The effort aimed to identify key themes from research with citizen designers and compare them to insights gained from two decades of research on banking and wealth using dominant UX methods. Initial analysis revealed themes that were further investigated, reinforcing the depth and variety of insights gained in self-directed conversations between designers and citizen designers. The themes that emerged were to do with Wealth, Just-ness (economic, value-based, and design value-based), Cultural Experiences, and Lived Experience. These are presented in the next subsections.

Wealth

The definition of wealth, interestingly, was not confined to "money." In fact, only four respondents felt money was the sole definition of wealth.

The definitions centered around the following:

- Health
- Freedom
- Family
- Education
- Ethics
- Satisfaction and Happiness

There were several interesting individual threads that were part of these conversations (Figures 1.1–1.5), such as the association of wealth with feudal decadence, quality of experiences rather than monetary value. A clear difference was observed in how some citizen designers perceived wealth as an individual matter, while others considered it in terms of the collective or social capital they were part of of/possessed.

Just-ness (Economic, Value-Based, and also Design Value-Based)

This theme was unexpected; I had unconsciously internalized the client-sponsored design ecosystem, where discussions on values like just-ness are often met with disdain. The argument is that citizens generally do not care about value-based just-ness, focusing instead on individual and material benefits, particularly economic just-ness when it directly benefits them. This perspective also aligns with two decades of user research I have conducted. Thus, the strong emergence of this theme of just-ness surprised me. Citizens' discussions about just-ness revealed significant awareness of inequalities, contradictions in value systems, and the need for balance. The insights centred around the following:

Research Cycle 1 – Whatsapp Conversation

When you think of the word "wealth," what is the first thing that comes to your mind...could be a word, a quote, an image, a story... anything that comes to your mind immediately.

FIGURE 1.1   Snapshots from the WhatsApp research.

Participant 9: Let me share a story. I was about 10 years old and on one of the weekends my family decided to visit a wealthy relative of mine who is a long travel from our house. Those times it took me took about two hours of travel on public bus and train as we didn't have a personal vehicle. On reaching our destination my father decided that since it was one of our first visit, we should buy a nice toy for the relative 's six-year-old son. When was past lunchtime we walked around the noon sun for 30-45 minutes searching for a Toyshop. Finally, we found this in Toyshop. The most amazing toys we went to the Lego section I was super excited because we did not go to toy shops such as these and I personally never had my toys. I don't really remember the price, but I distinctly remember it being very expensive. My father who is in a bank executive consider the decision for a while and decided to go ahead with the purchase. We reach the relative's house and I give the gift-wrapped toy to my dad's relative 's six-year-old child who had, during those times, had his own room. (He) runs out to break open the (gift) while the gift was still in my hands and check the notes. The moment he saw the toy he took the gift from my hand and threw it angrily to a corner of his room. He seemed annoyed with us since he already is to have a similar toy. Well back then toys were a special entity for me. I did not have too many of them and new one always were lot of waiting or a stellar achievement mostly. You know Grades. I was quite shocked that day to see this kind of behavior from a six-year-old. I remember his mother than took the new toy to a cupboard already filled with other such Lego boxes the reason I remember the story was while wealth just gives a person a sense of self-worth and independence, I wonder whether this comes at a price of constantly wanting to consume more and more. This I want and this behavior ends, and I wonder what a community becomes when everybody is only concerned about his or her own self-interest.

FIGURE 1.2   Snapshots from the WhatsApp research.

Participant 7: As a child, the word wealth meant jewellery. The colour yellow was associated with the word. As I would draw as a child and use yellow colour for jewellery. In college wealth meant property, bank balance. Now wealth means the ability to make my films. I mean wealth is the capacity to do whatever I want to do without any interference.

Participant 24: First thing that comes to my mind is having enough money which can fulfil my needs & my wishful needs/desires (expensive watches etc. etc.). Wealth should be enough to have a comfortable life... financial security... I believe when one is sad it's better to cry in a Mercedes than on a footpath.

FIGURE 1.3   Snapshots from the WhatsApp research.

1. Consensus on "decay of values"
2. Positive values emerging:
    1. Feminism/Women's rights
    2. Concerns about environment/sustainability
    3. LGBTQ inclusion/acceptance
    4. Rights of the marginalized

Participant 2:
First thing that comes is:
Wealth is long term value creation both in terms of money and personal ambitions/growth.

I feel the more wealth you have the more transactional your personal relationship becomes. But then for more wealth you have less available time so anyway we can't give time for worthwhile personal relationships.

As for materialistic personal ambitions I feel I don't know where to draw the line (like I want a car but is a BMW my limit). It's like more wealth I have, more show off might happen rather than fulfilling my utilitarian needs

FIGURE 1.4 Snapshots from the WhatsApp research.

नमस्ते माम,
कोहि धन (पैसा) शब्द सुनाते हो तो सबसे पहेले
मनमे ह जैसे एक शब्द आजाति उसेके बाद मन
सोचने लग्ति हे कि धन एक यैसा चिज हे कि धनसे
दुनिया मे अच्छा ओर बुरा दोही रस्ता देखादेंति है

अच्छा रस्ता:आधुनिकि करण ,,बिकास,
योजुकेसन,नये नये टेक्नोलोजि , बिलासि पन(बिदास
जिन्दगि) सुख सुबिधाये

बुरा रस्ता: घमन्ड,दुरब्यबाहार,
पारिवारिक प्रावलम,,प्राकिर्तिक (नेचर)प्रदुषण,
टेरोरिष्ट प्रोसाहित करना

FIGURE 1.5 Snapshots from the WhatsApp research.

3. Negative values emerging:

 1. Consumerism
 2. Individualism
 3. Loss of tolerance/compassion
 4. Overall, the world has become more inclusive and selfish/individualized at the same time.

Cultural Experiences

Understanding the citizen designers' cultural ecosystem and perceptions of wealth involved discussions of favorite books, poetry, music, films, and art. The influence of popular culture elements like Bollywood in music and films was clear. The books were mainly from the popular fiction genre, with some interesting exceptions. The question about sharing the image of a priceless and cherished art object that the citizen designer had at home elicited a variety of responses, ranging from old artifacts and photographs to artwork and artifacts created by the citizen designer themselves. Some examples of the descriptions accompanying the artefacts (Figures 1.6–1.9).

The responses highlighted a strong connection between memories and art objects. Cherishing something, it appeared, transcended monetary value, dispelling a myth in the business stakeholders' narrative. Banking stakeholders often believe in linking emotions in experience design to high-value "objects." They assume that happy memories must be tied to expensive items, like cars or holidays, justifying loans for these purchases.

Lived Experience

This theme reflects the citizen designers' experiences with and perspectives on two aspects of lived experience: education and news. When it came to education, there was a preponderance of thinking of education as cultural capital (good and bad memories of marks, grades, rank, prizes, recognition, etc.) on one hand, and a desire to shift away from traditional science and engineering-focused education to a more holistic and balanced approach on the other. It was as if there were ongoing inner conflicts between these two divergent paths.

Research Cycle 1 – Whatsapp Conversation

Can you share a photo of an art object at home that you cherish deeply… art object can be any sort of painting, figurines, masks, traditional artefacts like lamps etc. Why is it important?

FIGURE 1.6   Snapshots from the WhatsApp research.

I can't pin-point a reason as why is my fav painting BUT an idea I can associate with is that they were always there in the hall and my house is pretty small, so we had less artifacts but paintings as it hung on the wall. It sort of related to the idea of normalcy like home, if that makes any sense. When I look at these paintings I think about home. These paintings are not hung anymore because we had to paint the house. But we plan to hang them again because I love those paintings. They don't particularly mean anything, but it gives a homely feeling. I would call it as Hireath, it practically means longing for a house that's not there.

The paintings are hiraeth for me. I know my dad won't come back to Pune and my family can never go back what it was… like it was when I was a kid. So, these painting remind me of that time when everything was good.

Everything… was… home.

FIGURE 1.7   Snapshots from the WhatsApp research.

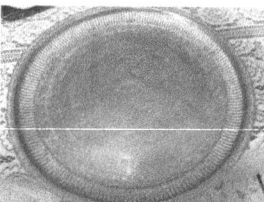

This is an object that I treasure. It is very beautiful and clearly old. I picked it up for 140 rupees from a shop that bought and sold old utensils in Bangalore's BVK Iyengar road in 1981. It is special for me partly because I got it so unexpectedly and so cheap. Also because it awakens a sadness deep within me when I think that it was definitely an article from a puja room and must have been in regular use for a long time, judging by how worn out the engraving was. I often wondered how such a beautiful plate ended up in the old utensil shop. Maybe it holds a lesson.

FIGURE 1.8   Snapshots from the WhatsApp research.

My Aai's photo – she made me who I am

FIGURE 1.9   Snapshots from the WhatsApp research.

Regarding news, there was considerable discontent about the state of the news media (this was when the Shashank Singh Rajput incident-related narratives and counter-narratives were prominent). Most citizen designers expressed unhappiness with the media's market orientation as well as the increasing restrictions on it.

Mentions of Black Lives Matter, American politics, and climate disasters in foreign countries came up in several conversations, while mentions of climate-related issues within the country or the Dalit movement were not as prominent. This raised questions about the sources of news for citizen designers. Most citizen designers in our group viewed online channels like BBC, CNN, and social media platforms as primary sources, supplemented by local newspapers (as the research was conducted during the first wave of the pandemic, newspapers were not readily available, relegating them to a secondary source) and TV.

Some Insights about Habitus, Capital, and Doxa

The research, despite its limited duration, provided a broad overview of citizen designers' habitus, capital, and doxa using Bourdieu's framework. Notably, it offered unique insights differing from previous studies through a research mode resembling an everyday WhatsApp conversation while employing Bourdieu's theoretical framework to guide the research direction and analysis. These insights highlight the value of further studies into these emerging areas of modalities.

Habitus (internalized social structures and dispositions that shape an individual's actions, beliefs, and tastes): A tug of war between individualistic and collective worldviews was evident throughout the WhatsApp conversations. Nonetheless, there was a general trend toward more expansive ways of defining one's identity and connection with the surrounding world.

On the question of wealth, the definition of wealth extended far beyond money and included various aspects, such as health, freedom, family, education, ethics, and happiness. This was also evident in the artifacts that signified value; these artifacts were not expensive objects but rather items that added meaning to a person's life in some way, even if they did not represent economic capital. Awareness and sensitivity regarding the issues of justice and discussions about the existence of inequalities, contradictions in value systems, and the need for balance also resonated throughout most conversations. The appreciation for the strengthening of women's positions in the world, the inclusion and emphasis on LGBTQ rights, movements against discrimination based on race, and concerns about the negative impacts of consumerism, nuclear families, and a self-centered attitude leading to a loss of compassion were also highlighted during the entire WhatsApp research. The market's grip on our lives was a concern for many citizens and emerged across topics, from health to education to the dissemination of news.

Capital (resources economic, social, cultural, and symbolic that individuals possess): Education as cultural and symbolic capital was very evident during the conversations in referring to grades, rank, prizes, recognition, etc. It was interesting to observe that viewing education as a means of accumulating social capital was not expressed. Cultural capital manifested in the form of art objects and artifacts that held personal significance for the citizen designers rather than monetary value.

Doxa (the taken-for-granted, unquestioned beliefs and values that are dominant in a particular social field): The WhatsApp conversations were able to highlight some of the contradictions present in people's worldviews. Let us consider the example of wealth and what it meant for our citizen designers.

Norms and values that are agreed upon/taken for granted—Wealth, for example, is about money, a big house, nice clothes, material comfort and security, health, freedom, privilege, power, investment, and financial planning.

Perspectives that are debated—Wealth has positive and negative aspects. It can make relationships transactional; promote conspicuous consumption, arrogance, and environmental destruction; and lead to corruption and familial conflict. Society, often remains unhappy despite material wealth due to an endless desire for more. Knowledge and regenerative systems, as well as spiritual and emotional well-being, quality friendships, community care, ethics, art, and culture, should also be valued as forms of wealth.

Regarding education, there was a conflict between the normative view of going with the flow, that of studying disciplines such as engineering, computer science, etc., which have high symbolic capital, versus an emergent desire to shift away from these traditionally valued disciplines and look for more holistic and balanced education.

EMERGENT QUESTIONS

The WhatsApp research and the use of Bourdieu's framework brought about new questions that need to be explored, such as:

1. What is the difference and connection between the habitus of a citizen designer who was a security guard and one who was a college professor, and how does this affect designed experiences? It may be tempting to reduce

habitus to quantifiable "things," objectifying citizens into "users" with certain economic and cultural capital. But does this objectification reflect the agency and potential of citizen designers, particularly the security guard? Does it reinforce inequitable social structures in design assumptions? What can be done to understand citizens' habitus more holistically and its role in design? Is the security guard's awareness of the duality of wealth perhaps an indicator of his agency to lead a life not totally dictated by the market?

2. Could the security guard and the college professor be potential citizen designers whose awareness of dualities and unequal social structures enables them to express ideas from their lived experiences that professional designers, observing through their expert lenses, at solving a design problem, might not be able to articulate?

3. Are the insights about the doxa (the "perspectives that are debated," for example) from the citizen designers, indicators of emerging perspectives often overlooked by UX researchers when employing traditional research methodologies?

The WhatsApp research with citizen designers had a significant impact on the relationship between the HFI designer (volunteer) and the citizen designer on the humanness of their interaction. This relationship was characterized by near equal participation and trust, making it more of a conversation than an interrogation. Additionally, it appeared to take place in a familiar environment of sharing worldviews and life experiences that largely evolved based on the citizen designers' inputs. The focus was on listening to and responding to the next question, even if it might not have been anticipated as a question of interest at the start of the research. The act of "listening," even if via WhatsApp, made the responses more personal and candid. The citizen designers felt engaged and wanted to share their stories rather than simply answering questions during a paid (and limited duration)session on a clearly defined topic.

THE DESIGN OF EVERYDAY SOLIDARITY INSTEAD OF EVERYDAY THINGS

Designers can empathize and connect with citizens in ways that corporate stakeholders cannot. They can leverage social and solidarity capital. Designers can "effect new connections and social relations that can alter ingrained structures" (Schalk, Kristiansson, and Maze 2017). Together with citizens, designers can co-create experience design co-ops, design collectives, and design commons. This new connection of solidarity should enable designers to stop viewing "users" and their understanding of the world as a mere source for projects of market value creation (Irani 2019). This shift will foster a movement for change in practice that does not reduce life to mere numbers and slots.

SOLIDARITY NETWORK WAS EFFECTIVE AND SUSTAINED

During the research cycle, a WhatsApp network enabled open conversations with citizen designers, allowing them to share viewpoints freely, unlike previous structured interviews or our ethnographic research in India. In later client-sponsored projects, our UX team aimed to replicate this WhatsApp-based method. In one program using WhatsApp, findings contradicted the MNC client's hypothesis. Traditional in-depth interviews via Zoom yielded more cautious responses from the paid 'participants' compared to those who responded via WhatsApp. This highlighted the stark differences in emotional richness between Zoom and WhatsApp responses. Our UX team felt compelled to challenge the client's views based on the WhatsApp responses that differed from the Zoom session responses.

After the research was completed, several connections on the WhatsApp network continued to be engaged in discussions on topics ranging from national headlines to popular culture and expressed curiosity about the previous research.

CONCLUSION

The WhatsApp research cycles highlight how WhatsApp-based UX research fosters more natural, empathetic, and sustained engagement, positioning participants as co-creators and enhancing solidarity capital for more inclusive design. Since completing the thesis, further research cycles using WhatsApp on various topics have been conducted, enabled by the UX4me platform, which combines WhatsApp with AI.

The UX4me platform shows that this approach can be scaled affordably while including citizens who are left out in traditional UX research because of where they are located or have difficulty accessing Zoom/Teams, etc. Further, a platform like UX4me enables the use of AI to efficiently manage and analyze the WhatsApp data, thus allowing UX researchers to look for meaningful insights instead of getting lost in the weeds. These findings challenge traditional market-driven research and support a UX ecosystem where citizens share equal stakes with designers and sponsors. Future research could compare WhatsApp with traditional methods across contexts, explore how participants' perspectives evolve over time, and study the impact of social roles on design contributions. It should also examine WhatsApp research in different languages and cultures and critically assess AI's role in data analysis, focusing on bias and context. Additionally, research could explore how to embed solidarity-based practices in design education and organizations and develop ethical frameworks for sustaining long-term participatory networks.

REFERENCES

Agamben, Giorgio. 2009. *What Is the Contemporary? What Is an Apparatus and Other Essays.* Stanford: Stanford University Press, Penguin Books.

Arendt, Hannah. 1963. *Eichmann in Jerusalem: A Report on the Banality of Evil.* New York: Penguin Classics.

Avgerou, Chrisanthi, Claudio Ciborra, and Frank Land. 2004. *The Social Study of Information and Communication Technology.* Oxford: Oxford University Press.

Bannon, Liam, and Pelle Ehn. 2013. "Design: Design Matters in Participatory Design." In *Routledge International Handbook of Participatory Design,* edited by Jesper Simonsen, and Toni Robertson, 37–63. New York: Routledge.

Bardzell, Shaowen. 2010. "Feminist HCI: Taking Stock and Outlining an Agenda for Design." In *Proceedings of the SIGCHI Conference on Human Factors in Computing*: 1301–10. New York: ACM.

Baumer, Eric P. S., and Jed R. Brubaker. 2017. "Post-Userism." In *Proceedings of the 2017 CHI Conference on Human Factors in Computing Systems—CHI 2017*: 6291–303. New York: ACM.

Bauwens, Michael, and Vasilis Niaros. 2017. "Changing Societies Through Urban Commons Transitions. P2P Foundation." Common Transition. Accessed Month Date, Year. https://commonstransition.org/wp-content/uploads/2017/12/Bauwens-Niaros-Urban-Commons-Transitions.pdf

Bodker, Susanne, Pelle Ehn, Dan Sjogren, and Yngve Sundblad. 2000. *Cooperative Design: Perspectives on 20 Years with 'the Scandinavian IT Design Model'*. Stockholm: CID.

Bourdieu, Pierre. 1971. "Intellectual Field and Creative Project." In *Knowledge and Control. New Directions for the Sociology of Education. Young*, edited by Michael F. D. Young: 89–119. London: Colleir-Macmillan.

Bourdieu, Pierre. 1977. *Outline of a Theory of Practice*. Cambridge, MA: Cambridge University Press.

Bourdieu, Pierre. 1984. *Distinction: A Social Critique of the Judgement of Taste*. Cambridge, MA: Harvard University Press.

Bourdieu, Pierre. 1986. "The Forms of Capital." In *The Handbook of Theory and Research for the Sociology of Education*, edited by John G. Richardson: 241–58. New York, NY: Greenwood Press.

Bourdieu, Pierre, and Loïc J. Wacquant. 1992. *An Invitation to Reflexive Sociology*. Chicago: University of Chicago Press.

Calhoun, Craig, Edward LiPuma, and Moishe Postone, eds. 1993. *Bourdieu: Critical Perspectives*. Chicago: University of Chicago Press.

Chapanis, Al 1979. "Quo Vadis Ergonomia." *Ergonomics* 22 (6): 595–605. doi:10.1080/00140137908924644

Clarke, Victoria, Virginia Braun, and Nikki Hayfield. "Thematic Analysis." In *Qualitative Psychology: A Practical Guide to Research Methods*. 3rd ed. vol. 2015, edited by J. A. Smith. London: SAGE.

Creative Reaction Lab. n.d. "A new model for Community Engagement, Problem Solving, and Creating Inclusive and Equitable Outcomes." Creative Reaction Lab. Accessed. https://www.creativereactionlab.com/our-approach

Dunne, Anthony. 1999. *Hertzian Tales: Electronic Products, Aesthetic Experience, and Critical Design*. Cambridge, MA: MIT Press.

Ehn, Pelle, and Morten Kyng. 1987. "The Collective Resource Approach to Systems Design." In *Computers and Democracy- A Scandinavian Challenge*, edited by Gro Bjerknes, Pelle Ehn, and Morten Kyng: 17–58. Aldershot: Gower Publishing Ltd.

Escoffier, Simón. 2018. "Mobilisational Citizenship: Sustainable Collective Action in Underprivileged Urban Chile." *Citizenship Studies* 22 (7): 769–90. doi:10.1080/1362 1025.2018.1508412

Esteves, Ana Margarida. 2019. "Constructing Solidarity Economy-Based Livelihoods." *Museum International* 71 (1–2): 132–9. doi: 10.1080/13500775.2019.1638069

Forno, Francesca, Cristina Grasseni, and Silvana Signori. 2015. "Italy's Solidarity Purchase Groups as 'Citizenship Labs.'." In *Putting Sustainability into Practice*, edited by Emily H. Kennedy, Maurie J. Cohen, and Naomi Krogman, 67–88. Broad Heath: Edward Elgar Publishing.

Foucault, Michel. 1984. *The Foucault Reader, edited by Paul Rabinow*. New York: Pantheon Books.

Ghaffarian, Vafa. 2011. "The New Stream of Socio-technical Approach and Mainstream." *Procedia Computer Science* 3: 1499–511. doi:10.1016/j.procs.2011.01.039

Gibbs, Graham R. 2012. *Analysing Qualitative Data. The Sage Qualitative Research Kit*. London: SAGE.

Gil de Zúñiga, H., A. Ardèvol-Abreu, and A. Casero-Ripollés. 2021. "WhatsApp Political Discussion, Conventional Participation and Activism: Exploring Direct, Indirect and Generational Effects." *Information, Communication & Society* 24 (2): 201–18. doi:10.1 080/1369118X.2019.1642933

Harker, Richard, Cheleen Mahar, and Chris Wilkes, eds. 1990. *An Introduction to the Work of Pierre Bourdieu.* London: Macmillan.

Hirschheim, Rudy, and Heinz K. Klein. 1994. "Realizing Emancipatory Principles in Information Systems Development: The Case for Ethics." *MIS Quarterly* 18 (1): 83–109. doi:10.2307/249611

Ignatow, Gabe, and Laura Robinson. 2017. "Pierre Bourdieu: Theorizing the Digital." *Information, Communication & Society* 20 (7): 950–66. doi:10.1080/1369118X.2017. 1301519

Irani, Lilly. 2019. *Chasing Innovation: Making Entrepreneurial Citizens in Modern India.* Princeton, NJ: Princeton University Press.

Kraus, C. 2014. "Inside Goal – Directed Design: A Two-Part Conversation with Alan Cooper." Accessed April 2015 https://www.cooper.com/journal/2014/04/inside-goal-directed-design-a-two-part-conversation-with-alan-cooper

Kuutti, Kari. 1996. "Activity Theory as a Potential Framework for Human-Computer Interaction Research." In *Context and Consciousness: Activity Theory and Human-Computer Interaction,* edited by Bonnie A. Nardi, 17–44. Cambridge, MA: MIT Press.

Laville, Jean-Louis. 2016. *A Economia Social e Solidária: Práticas, Teorias e Debates.* Coimbra: Almedina.

Macey, David. 2000. *The Penguin Dictionary of Critical Theory.* London: Penguin Books.

Manzini, Ezio. 2015. *Design, When Everybody Designs: An Introduction to Design for Social Innovation.* Cambridge, MA: MIT Press.

Mumford, Lewis. 1923. *Technics and Civilization.* London: Routledge and Kegan Paul Ltd.

Myers, Michael D. 1997. "Critical Ethnography in Information Systems." In *Information Systems and Qualitative Research,* edited by A. S. Lee, J. Liebenau, and J. I. DeGross, 276–300. London: Chapman & Hall.

Myers, Michael D., and Heinz K. Klein. 2011. "A Set of Principles for Conducting Critical Research in Information Systems." *MIS Quarterly* 35 (1): 17–36. doi:10.2307/23043487

Norman, Don. 2005. "Human-Cantered Design Considered Harmful." Accessed December 2014 https://www.jnd.org/dn.mss/human-centered_design_considered_harmful.html

Orlikowski, Wanda J., and Jack J. Baroudi. 1991. "Studying Information Technology in Organizations: Research Approaches and Assumptions." *Information Systems Research* 2 (1): 1–28. doi:10.1287/isre.2.1.1

Papanek, Victor. 1985. *Designing for the Real World: Human Ecology and Social Change.* Chicago, IL: Academy Chicago Publishers.

Pareto, Vilfredo. 1971. *Manual of Political Economy.* Translated by Ann S. Schweir & Alfred N. Page. New York: A.M. Kelley.

Rakopoulos, T. 2015. "Solidarity Economy in Contemporary Greece: 'Movementality', Economic Democracy and Social Reproduction During Crisis." In *Economy for and Against Democracy,* edited by Keith Hart: 161–81. London: Berghahn Book Company.

Rheinfrank, John. 2002. "The Philosophy of (User) Experience." Presentation at *CHI2002/AIGA Experience Design Forum,* Minneapolis, April 2002.

Rittel, Horst W. J., and Melvin M. Webber. 1973. "Dilemmas in a General Theory of Planning." *Policy Sciences* 4 (2): 155–69. doi:10.1007/BF01405730

Saldaña, Johnny. 2014. "Coding and Analysis Strategies." In *The Oxford Handbook of Qualitative Research,* edited by Patricia Leavy. Oxford: Oxford University Press.

Sanders, Elizabeth B. 2002. "From User-Centered to Participatory Design Approaches." In *Design and the Social Sciences,* edited by Jorge Frascara, page range. London: CRC Press.

Schalk, Meike, Thérèse Kristiansson, and Ramia Maze, eds. 2017. *Feminist Futures of Spatial Practice: Materialisms, Activisms, Dialogues, Pedagogies, Projections*. Baunach: Spurbuchverlag.

Schon, Donald A. 1959. *The Reflective Practitioner*. New York: Basic Books.

Simonsen, Jesper, and Toni Robertson, eds. 2013. *Routledge International Handbook of Participatory Design*. Abingdon: Routledge.

Stahl, Bernard Carsten, and Carole Brooke. 2008. "The Contribution of Critical IS Research." *Communications of the ACM* 51 (3): 51–5. doi:10.1145/1325555.1325566

Thomas, Jim. 1993. *Doing Critical Ethnography*. Newbury Park, CA: SAGE.

Tonkinwise, Cameron. 2013. "Design Away: Unmaking Things." Academia.edu. Accessed Month Date, Year. https://www.academia.edu/3794815/Design_Away_Unmaking_Things

Treré, E. 2020. "The Banality of WhatsApp: On the Everyday Politics of Backstage Activism in Mexico and Spain." *First Monday*. doi:10.5210/fm.v25i12.10404

2 Culturally Appropriate Responsible Design (CARD)

Girish Prabhu

INTRODUCTION

User Experience Design helps designers in creating functionally and aesthetically pleasing interactions for users around the world. User Experience Design today is focused on customer acquisition and retention. Such an approach creates an experience economy that relies on instant gratification. In order to save humanity, it is important to create a sustainable lifestyle that promotes responsible consumerism (Lahiri Chavan and Prabhu, 2020).

For that, we need a conceptual framework that considers factors important in designing a solution, a framework that would push design research ahead and may improve results for the users, community and society. A comprehensive framework for UX design will allow the designers to select the correct design strategies based on a deeper understanding of the design problem. Researchers and Designers could benefit from methodological frameworks that address broader concerns related to ethics, inclusivity, environmental impact, long-term sustainability and diversity of cultures.

I have drawn upon two conceptual frameworks – a framework for cultural design and the other from responsible design superimposed on the standard human-centred design framework to develop the Culturally Appropriate Responsible Design (CARD) framework. For cultural design, I include the trichotomy of needs, values and capabilities at the individual and the family level. For responsible design, similarly, I consider societal, economic and environmental impact (see Figure 2.3).

BACKGROUND

In addition to focusing on the needs of individual users, it's crucial to consider families, communities and societies as a whole. Technology plays a key role in most of the systems now, and all these elements should be taken into account during the design process. In this chapter, I encourage the development of mindsets that will allow designers to consider, in a systematic manner through critical research & design, a range of alternative sociocultural-technical design options.

DOI: 10.1201/9781003557777-3

LIMITATIONS OF TRADITIONAL HUMAN-CENTRED DESIGN APPROACH

According to the International Organisation for Standardisation standard, human-centred design focuses on the users' needs and requirements and promises to make systems, products and services useful and usable. Especially, when it comes to socio-cultural design, human-centred design has limitations on what it can accomplish.

Most Human Centred Design (HCD) methods are hampered by the fact that human-centred design focuses on the individual user, their needs and expectations. Within this micro-context, the social, ethical and interpersonal issues are missed. It does not provide critical consideration to promote human interaction, empathy, encourage a shared sense of purpose, community building and convergence that feel culturally appropriate and responsible. These issues play an ever-important role in any responsible design.

The Double Diamond process of design thinking, developed by the British Design Council in 2005, showcases aspects of design that are very important. This process allows a designer to focus both on what and how questions (solutions) and also the why questions (problems). However, the process considers the definition of a problem and the final solution design as two separate parts of the design. It also assumes a linear phase-wise progression through the four phases of Discover, Define, Develop and Deliver. In reality, most design projects that are successful are never linear and go back & forth between the four phases. Most successful design projects are also very iterative. The Double Diamond misses these key design elements.

PHILOSOPHY BEHIND CULTURALLY APPROPRIATE RESPONSIBLE DESIGN

To be able to create solutions that are responsible and also culturally appropriate, the traditional double-diamond human-centred design thinking process needs to be modified. On the theoretical and design side of this equation, the designer needs to apply the principles of design thinking and human-centred design at every stage, namely, Explore > Define > Conceive > Evolve, as shown in Figure 2.1. In other words, the problem definition and solution design stages are not separate; they are intertwined. For example, at the Explore stage, one starts defining, conceiving and evolving a solution, and at the same time questions the problem and solution critically in a very non-linear way using convergent–divergent thinking and applying both critical and creative thinking.

As the designer moves through this process iteratively, it is important for the designer to keep applying the CARD Framework through critical research and the design process. This approach allows designers to go beyond the individualistic human wants. It covers human needs, values and capabilities-based on family and community. And further expands it to include society, economy and environment.

CRITICAL RESEARCH

Critical research and design has gained popularity since the 1960s. It pushes designers to deliver design products that are more conscious by forcing them to think differently. A vast majority of designers create affirmative design, contributing indirectly

Explore
Understanding what exists today, what has driven the "present" & will drive the future, and what the stakeholders envision. Creation of boundaries within which the solution will be thought of, conceived and formulated.

Define
Identifying opportunities in line with the goals and capabilities of the organization and the market and ecosystem needs and limitations. Opportunities are based on:

- Vision, business and capability
- Positioning of the competitors
- Opportunities within the market

Conceive
Formulation of ideas and developing them into proof of concepts

Evolve
Iteratively redefining and redesigning the concepts based on feedback from the users and relevant stakeholders

Critical Thinking

Explore *Define* *Conceive* *Evolve*

Creative Thinking

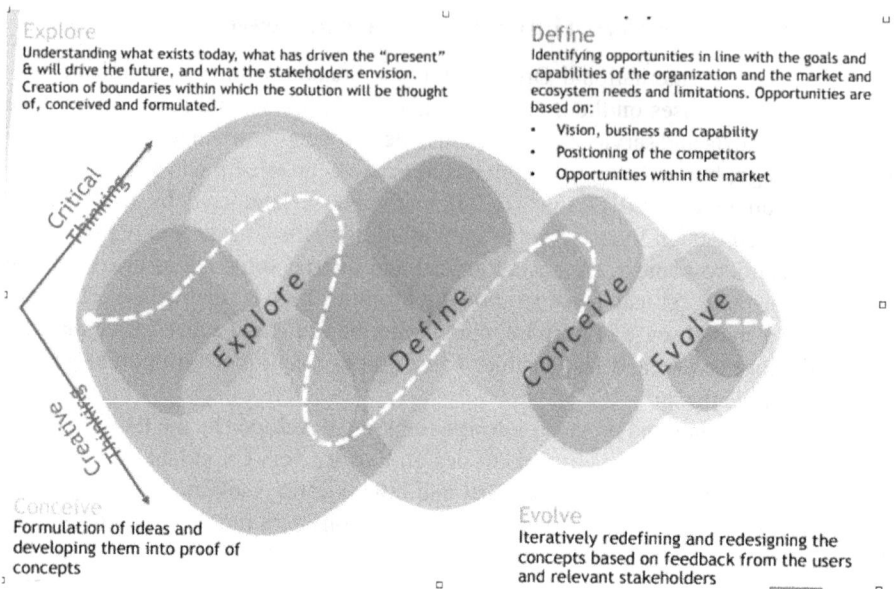

FIGURE 2.1   Modified double diamond human-centred design process.

to global issues with negative social and environmental implications (Dunne & Raby, 2001). Critical design approach focuses on probable futures using critical analysis and delivers benefits by increasing the designer's ideology.

Influenced by the critical social theory, critical design centres on "present social, cultural and ethical implications of design objects and practice" (Malpass, 2013). The designer's role and responsibility in making the users aware of their involvement as citizens is emphasised by critical design. In doing so, the designer is allowed to depart from the affirmative design process and use alternative frameworks such as the CARD framework instead of "being informed by values based on a specific world view, way of seeing and understanding reality" (Dunne & Raby, 2001).

While using the CARD framework, designers have to utilise critical research principles to gather insights and reflect on them throughout the Evolve–Define–Conceive–Evolve iterative double diamond process. Drawing upon the principles of interpretive research and critical research, designers can incorporate one or more of the following principles of generating insights, critique and creating transformation in their work (Myers & Klein, 2011).

Principles to Generate Insights

1. Hermeneutic Circle Principle – Develop understanding by considering the meaning of parts and the whole that they form.
2. Contextualisation Principle – Reflect on the historical and social background.
3. Interaction Principle – Observe how the research findings were constructed through the interaction between the users and researchers.

4. Abstraction and Generalisation Principle – Describe the unique details uncovered by the interpretation of concepts that describe the human understanding and action.
5. Dialogical Reasoning Principle – Adapt your understanding accordingly when the designers find contradictions between their preconceived notions and what the data actually reveals.
6. Multiple Interpretations Principle – Be open towards differences in interpretations among the players and stakeholders.
7. Suspicion Principle – Ensuring the validity and reliability of qualitative research by being open to biases and differences in participant narratives.

Principles to Critique

1. Core Concepts Principle – Utilise concepts and ideas from critical social theorists to arrange your data collection and analysis.
2. Value Position Principle – Support values such as discursive ethics, open democracy or equal opportunity.
3. Challenging Prevailing Beliefs/Social Practices Principle – Challenge beliefs and practices with conflicting arguments and evidence.

Principles to Create Transformation

1. Individual Emancipation Principle – Promote the fulfilment of human needs and possible self-reflection and self-transformation.
2. Improvements in Society Principle – Suggest how social improvements are possible without assuming any authoritative position.

CRITICAL DESIGN

Apart from the existing design principles, based on critical theory approaches of Bourdieu, Foucault and Habermas, I have devised the following principles of equitable design:

1. Distribution of capitals: Design to provide appropriate access to economic, social, cultural and symbolic capitals within the social space that each individual occupies due to their social conditioning.
2. Access to knowledge: Provide access to the right amount of knowledge, as knowledge is power
3. Self-surveillance and self-awareness: Provide various mechanisms that regulate the behaviour of individuals, using a combination of self-surveillance and self-awareness approaches.
4. Individual and mutual success: Allow cooperation through mutual understanding, and a correct level of competition based on their own private goals
5. Bridge the gap between internal and external world: Bridge the gap between the internal subjective viewpoint of the lifeworld, experienced due to culture, society or personality and the external systems.

In the critical design process, I also focus on the following four pillars of R.I.S.E. (explained below and in the following sections with an example of John Doe's personal automobile usage.)

1. Responsible Future: Think about a preferred future rather than a possible and probable future, combining futures methods and speculative/critical design methods. How do we understand and modify John's behaviour towards public transportation or smaller cars? Should we build more fuel-efficient cars? Should we design policies that promote public transportation?

2. Impact Analysis: Continuously think about the impact your decisions can have on all the ecosystem stakeholders. How does John's decision to use his personal automobile impact current cab owners and how does it impact the general car manufacturing industry? What impact does it have on the current car drivers, rickshaw drivers? What are the other impacted groups?

3. Systems Thinking: Think about the interventions as a connected network and not just a product. Should we invest in infrastructure and policies that promote public transportation? How do we provide last-mile connectivity for John?

4. Ecosystem Alignment: One needs to go beyond the human-centred design approach of aligning only the users and business goals. Think about your community, society and the ecology. In short, use a humanity-centred design. How do I redesign my infrastructure intervention so that it does not split the communities living in the area through which the new roads/metro/high-speed railways would pass? How do I reduce the impact on the environment and ecology?

R.I.S.E. allows responsible and equitable design and innovation for a bigger impact. This approach not only explores desirability, feasibility and viability but also focuses on societal/ecological impact. It allows designers to think about the impact and cost of their interventions. It works both at the strategic and the tactical level. At the strategic level, designers can think about the positive impact, the global and social responsibility, culture, values, ethics and justice. At the tactical level, apart from aesthetics, usability, accessibility, designers need to think about socio-cultural-technical issues, behavioural impact, materials, environmental impact and responsible business models. In short, R.I.S.E. allows us to strive to improve our understanding of the impact of our work.

Methods and tools such as futures and strategic forecasting, systems mapping, speculative design, persuasive and emotional design and cooperative business model innovation, to name a few, already exist and can be used to implement this approach. It just requires a mindset change.

CULTURALLY APPROPRIATE RESPONSIBLE DESIGN FRAMEWORK

The CARD framework explains factors relevant to sociocultural aspects of design. Even though the solution is presented as relevant to an individual usage case, that is aligned with the needs, values and capabilities of that individual, the designer gives appropriate consideration to the cultural design contexts such as family and community, and the responsible design contexts of economy and environment. As society can have an impact on the usage and vice versa, the framework also emphasises that the influence of society is extremely important to consider.

The CARD framework can be used to evaluate and manage the general impact and equity of the design. Design researchers can then adapt certain elements of the design to reduce overall prejudice. We encourage designers to use this framework for socio-cultural-technical problems. This framework better includes ethics, fairness, inclusivity, environmental impact, long-term sustainability and diversity of cultures during a design effort.

CULTURAL ASPECTS OF THE FRAMEWORK

On the cultural aspects of design, the CARD framework includes needs, values and capabilities of users at various levels – the individual user, their family and their community level. To understand the needs of participants (in contrast to just wants and desires), I utilise Manfred-MaxNeef's fundamental human needs model that lists Subsistence, Protection, Affection (or love), Understanding, Participation, Idleness, Creation, Identity and Freedom as the key universal human needs (Max-Neef, 2017). I posit that these needs have to be juxtaposed with the wants and desires of the participants at the above-mentioned levels of the individual user, their family and their community.

Similarly, the framework posits that the Values of the participants should be explored based on the theory of human universal values, a theory developed by Shalom H. Schwartz (Schwartz, 2012). This theory expands on the cross-cultural communication framework proposed by Hofstede in his cultural dimensions theory (Hofstede, 2011). Schwartz has identified eleven basic human values that cover independent thought, autonomous action, excitement in life, pleasure, success, social status, safety, behavioural restraint, respect for customs, preserving and enhancing the welfare of all people, welfare of nature and spirituality. Each of these 11 values identified by Schwartz also varies across different cultures and across the axes of the individual, their family and the community they are associated with.

The third cultural aspect is that of human capability. Nussbaum (2000) proposes 10 capabilities. These are opportunities based on social and personal circumstances. According to Nussbaum, even within a family unit, individuals can have greatly different needs (Nussbaum, 2011). Nussbaum's argument is to develop capabilities at a personal level rather than at a group level. The capabilities that need to be supported are: (1) Ability to reach the end of a human lifespan; (2) Ability to maintain good health; (3) Having freedom to travel from one location to another; (4) Ability to perceive through the senses, imagine, think and reason; (5) Ability to form connections with things and people beyond oneself; (6) Ability to develop a personal understanding of what is good and to engage in thoughtful reflection on life's planning; (7) Ability to coexist and build relationships with others while having the social foundations for self-respect and dignity; (8) Ability to live with consideration for and in harmony with animals, plants and the natural environment; (9) Ability to laugh, engage in play and take pleasure in recreational activities; and (10) Ability to exert influence over one's surroundings, both politically and materially.

RESPONSIBILITY-RELATED ASPECTS OF THE FRAMEWORK

Interestingly, all three cultural aspects (needs, values and capabilities) of the framework capabilities cannot be seen in isolation from sustainable and responsible design. Hence, the framework posits that each of these has to be critically evaluated to understand and optimise the impact on the environment, economy and society.

To continue with John Doe's need for mobility, Holland (2008) suggests that imposing mandatory restrictions on the use of Sports Utility Vehicles (SUVs), for instance, serves as an indirect method of setting capability limits. This would free up resources that could be reallocated to support essential entitlements, like enhancing health. While such a restriction would enhance the capabilities of others, it would nonetheless impose a limit on the freedom of movement for John Doe, who wants to drive SUVs. Consider John, who lives close to the centre of a large city, as an example. Suppose he has several options to exercise his mobility capability and access the city centre: he can walk, ride a bicycle, use public transportation or drive his SUV. Prohibiting the most environmentally harmful option (i.e. driving an SUV) would still allow John to exercise his mobility capability, as he could walk or use public transportation to reach the city centre. However, such a measure would breach the principle of liberal neutrality concerning different conceptions of the good life, thereby disregarding John's fundamental interest in selecting his preferred way of life.

Alternatively, Peeters et al. (2015) suggest a functioning threshold or functioning constraints. In this scenario, to operationalise constraints on functioning, the suggestion is to lower emissions by implementing personal carbon allowances (PCAs), which would be allocated based on an individual's share of the total allowable emissions budget (Hyams, 2009). To achieve a gradual decrease in greenhouse gas emissions, the total number of PCAs issued each year would be reduced. Consequently, to participate in activities that emit greenhouse gases, individuals would need to present their PCAs. In John's case, he would be permitted to drive an SUV, but it would consume most or all of his carbon allowance. He might choose to forgo other activities (such as giving up trans-continental vacations or cutting back on meat and dairy consumption) or decide to drive his SUV only occasionally to allocate more of his budget towards driving. Additionally, he could potentially offset the greenhouse gas emissions from driving by investing in renewable energy or enhancing carbon sinks.

The CARD framework thus allows designers to evaluate various alternative proposals to develop an optimal, sustainable and responsible design.

RESEARCH AND DESIGN PROCESS

As discussed in section "Background", the CARD framework (Figure 2.2) is applied within the modified double diamond human-centred design process (Figure 2.1). The schematic in Figure 2.3 shows the overall integrated process.

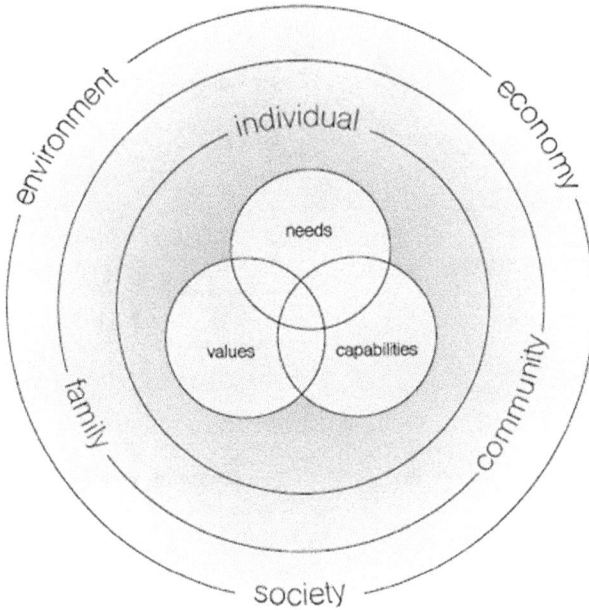

FIGURE 2.2 Culturally appropriate responsible design (CARD) framework.

In the following section, I have made an attempt to apply the overall framework using a fictitious use case. To continue with John Doe's mobility usage requirement, one would start with the modified double diamond method to understand a person like John's needs, values and capabilities. Throughout the four stages of iterative design (i.e. Explore ↔ Define ↔ Conceive ↔ Evolve), critical research theories and one or more critical research principles of generating insights, critique and creating transformation and principles of equitable design are applied.

As shown in the schematic in Figure 2.3, along with that, the designer uses the four key areas of questions, namely, how to create a responsible future, how to create an impact, how to think about the whole system and how to develop ecosystem alignment, as shown in Figure 2.4.

During the Explore stage, it is important to understand John's perspective of the impact of his needs, values and capabilities on his family, community, society, economy and environment. During this cultural research stage, the designer engages with people similar to John's persona and makes an attempt to understand how the data collected aligns or deviates from the three models, namely (1) Manfred-MaxNeef's fundamental human needs model, (2) Schwartz's human universal values model and (3) Nussbaum's capability model.

For example, after the cultural research, let us assume that the researcher compiled following needs for John's needs: (a) wants physical and mental health, (b) desires to have care, adaptability and autonomy, (c) cares about relation with friends, (d) wants to relax and have fun, (e) wants sense of belonging, self-esteem and (f) needs autonomy. Similarly, let us assume that John's values include: (a) social status, (b) pursuit

Culturally Appropriate Responsible Design Framework (CARD)

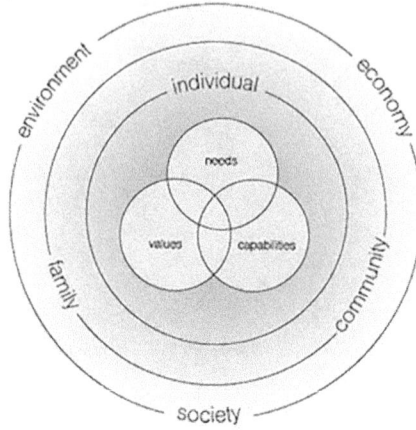

environment

individual

economy

needs

values   capabilities

family

community

society

Explore <—> Define <—> Conceive <—> Evolve

Human Centred Design (HCD)

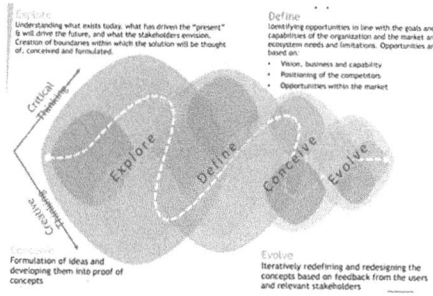

Explore
Understanding what exists today, what has driven the "present" & will drive the future, and what the stakeholders envision. Creation of boundaries within which the solution will be thought of, conceived and formulated.

Define
Identifying opportunities in line with the goals and capabilities of the organization and the market and ecosystem needs and limitations. Opportunities are based on:
• Vision, business and capability
• Positioning of the competitors
• Opportunities within the market

Critical Thinking

Explore   Define   Conceive   Evolve

Creative Thinking

Conceive
Formulation of ideas and developing them into proof of concepts

Evolve
Iteratively redefining and redesigning the concepts based on feedback from the users and relevant stakeholders

Critical Research and Design Principles

Cultural Research

Principles to Generate Insights
• Hermeneutic Circle
• Contextualisation
• Interaction
• Abstraction and Generalisation
• Dialogical Reasoning
• Multiple Interpretations
• Suspicion

Principles to Critique
• Core Concepts
• Value Position
• Challenging Prevailing Beliefs / Social Practices

Responsible Design

Principles to Create Transformation
• Individual Emancipation
• Improvements in Society

Principles of Equitable Design
• Distribution of capitals
• Access to knowledge
• Self-surveillance and self-awareness
• Individual and mutual success
• Bridge the gap between internal and external world

RISE

Responsible Future Impact Analysis

Systems Thinking Ecosystem Alignment

FIGURE 2.3 Research and design methodology.

FIGURE 2.4　Explore–define–conceive–evolve.

						Needs/Values/Capabilities emerged after Critical Research Probing					Needs/Values/Capabilities either not important or relevant				
Explore	Legend: –>	subsistence	protection	affection	understanding	participation	leisure	creation	identity	freedom	security
	Needs (MaxNeef)	Needs physical and mental health	Needs care, adaptability and autonomy	Cares about relation with friends, with nature	Has curiosity, intuition		Wants to relax, have fun		Wants sense of belonging, self-esteem	Needs autonomy (but is flexible)	Cares about safety and stability of society
	Values (Shwartz)	Power	Achievement	Hedonism	Stimulation	Self-Direction	Universalism	Benevolence	Tradition	Conformity	Security
		Values social status	Values pursuit of success	Values pleasure, enjoyment	Seeks excitement, variety and adventure		Appreciates the welfare of all people and nature		Accepts the ideas that traditional culture provides		Cares about safety and stability of society
	Capabilities (Nussbaum)	Ability to reach the end of a human lifespan	Ability to maintain good health	Having freedom to travel	Ability to perceive through the senses	Ability to form connections with people	Ability to develop an understanding of good	Ability to build relationships with others	Ability to live in harmony with ecology	Ability to laugh, and engage in play	Ability to influence one's surroundings
		Relevant		Relevant		Relevant			Relevant		
Define	Insights 1		Cares about relation with friends, with nature	Appreciates the welfare of all people and nature	Ability to live in harmony with ecology		4 Needs autonomy (but is flexible)	Accepts the ideas that traditional culture provides	Ability to form connections with people		
	2		Needs physical and mental health	Cares about safety and stability of society	Ability to reach the end of a human lifespan						
	3		Wants sense of belonging, self-esteem	Values social status	Ability to live in harmony with ecology						
	4		Needs autonomy (but is flexible)	Accepts the ideas that traditional culture provides	Apply to form connections with people						
Conceive	Responsible Solutions 1	Safe, sturdy, vehicle for self ownership that promotes social status (From Insights 2 and 3)					4 Programs such as personal carbon allowances (PCAs) (From Insight 3)				
	2	Safe safe of public transportation with last mile connectivity that adds to the social status (From Insights 2 and 4)					5				
	3	Social status based shared services so the people in the ecosystem (From Insight 4)					6				

1	Cares about relation with friends, with nature	Appreciates the welfare of all people and nature	Ability to live in harmony with ecology
2	Needs physical and mental health	Cares about safety and stability of society	Ability to reach the end of a human lifespan
3	Wants sense of belonging, self-esteem	Values social status	Ability to live in harmony with ecology
4	Needs autonomy (but is flexible)	Accepts the ideas that traditional culture provides	Ability to form connections with people

FIGURE 2.5   Sample insights generated through critical research and design.

of success, (c) pleasure & enjoyment, (d) seeks excitement, variety and adventure, (e) accepts the ideas that traditional culture provides and (f) cares about safety and stability of self. The capabilities John wants to have are (a) the ability to live long, (b) the ability to maintain good health, (c) has freedom to travel where ever he wants and (d) the ability to form connections with people.

Throughout the cultural research during the evolve stage, any differences and deviations that the researcher notices from the three models for needs, values and capabilities are evaluated and probed using the critical research and design principles mentioned in Figure 2.3 and elaborated in section "Critical Research" to generate optimal insights. This process uncovers following additional needs, values and capabilities: (a) Has curiosity and intuition of what is good for him and the environment, (b) Cares about his relation with nature/ecology, (c) Needs autonomy *but is willing to be flexible*, (d) Appreciates the welfare of all people and nature, (e) Cares about safety and stability of society and (f) Needs to be able to live in harmony with ecology. As one can notice, apart from new needs, the need to have autonomy has changed to "*needs autonomy but is willing to be flexible*" after this round of probing. Utilizing the principles of critical research and design allows the researcher to make the users realise the importance of culturally appropriate and responsible design.

During the Define stage of the modified double diamond process (Figure 2.1 in section "Philosophy behind Culturally Appropriate Responsible Design"), the designer evaluates these needs, values and capabilities to generate coherent insights as shown in Figure 2.5 (note: grey boxes indicate insights generated after probing).

During the Conceive stage of the modified double diamond process, the designer develops ideas using various design methods such as ideation, participatory design and speculative design, generating ideas that may appeal to John. The designer also utilises the principles of equitable design and the four pillars of R.I.S.E. (as explained in section "Philosophy behind Culturally Appropriate Responsible Design")

- Responsible Future: How can the concept modify John's behaviour towards public transportation or smaller cars? How to design more fuel-efficient cars? What policies and interventions are required that will promote his usage of public transportation?

- Impact Analysis: What interventions are required to mitigate the impact of John's decision to use his personal automobile on current cab owners, rickshaw drivers and other impacted groups? How does it impact the general car manufacturing industry?
- Systems Thinking: Should we invest in infrastructure and policies that promote public transportation? How do we provide last-mile connectivity for John?
- Ecosystem Alignment: How do I reduce the impact on the environment and ecology?

Post this exercise, the sample ideas could be as follows:

- Safe, sturdy, vehicle for self-ownership that promotes social status
- Safe and easy to use public transportation with last-mile connectivity that adds to the social status
- Social status based shared services so that cab drivers, rickshaw drivers and other people in the ecosystem benefit
- Programs such as personal carbon allowances (Peeters et al., 2015 mentioned in section "Responsibility-Related Aspects of the Framework") to provide autonomy with flexibility

And finally, during the Evolve stage of the modified double diamond process, the designer probes John further to evaluate and tweak the ideas. It has to be noted that the designer (and researcher) moves back and forth through these four stages of iterative design (i.e. Explore ↔ Define ↔ Conceive ↔ Evolve)

CONCLUSION

As narrated in the section above, the CARD framework and associated methodology not only allow the designer to develop responsible design solutions but also allow the participants to become more aware of their role in being a responsible citizen. As shown in Figure 2.6, the impact of this approach will be felt when the designs generated using the "Theory & Design" side are actually implemented and experienced in the real world.

We, I and my colleagues at Srishti & HFI, have applied one or more elements of the CARD framework to two design challenges, namely, (a) AutoRaja – creating livelihood for the bottom of the pyramid auto-rickshaw drivers in India (Prabhu and Saraf, 2015) and (b) Public Spirit – Self Worth (HSI) instead of Net Worth (HNI) – A Speculative Design exploration, envisioning the future of money in an open access economy (Prabhu et al., 2022). In the AutoRaja project, we explored the challenges of Indian Auto Rickshaw drivers' community, as the sustainability of this community is threatened by the numerous institutional and socio-cultural problems that the drivers face, while the Indian Auto Rickshaw continues to provide the Indian commuter with the much-needed last-mile connectivity. Designers envisioned using technology as an enabler to create a self-regulated, self-reliant and economically empowered community that will provide urban citizens a safe and reliable mode of transport. In the Public Spirit project, the designers explored whether social self-worth can replace

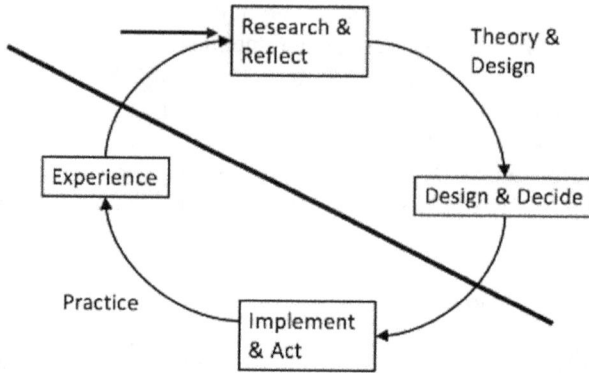

FIGURE 2.6   Theory and design <——-> practice.

the financial net worth of an individual and a community, and developed a mock-up to show how daily transactions can be done using self-worth token instead of money.

ACKNOWLEDGEMENT

This chapter was developed based on my work experience at Srishti Labs at Srishti Manipal Institute of Art, Design & Technology and at HFI.org.in. I thank Atul Saraf, Beena Prabhu and Apala Lahiri for numerous discussions that led to this thought process. The AutoRaja project of creating livelihood for bottom of the pyramid auto rickshaw drivers in India was led by Beena Prabhu & Atul Saraf at Srishti labs, and I thank them for their contributions to this chapter. Beena was invited to present this work at the Smart City Expo World Congress in Barcelona in 2015. The Public Spirit project was done at HFI.org.in and I thank our interns Shraeyas Massey and Purva Dalvi for their efforts in applying parts of the CARD framework on this project. Last but not least, I thank Rahul Prabhu for his efforts in creating all the figures for this chapter and also for proofreading and editing this chapter.

REFERENCES

Dunne, A. and Raby, F. (2001), *Design Noir: The Secret Life of Electronic Objects*. Springer Science & Business Media, N/A. Google Scholar.

Hofstede, G. (2011). Dimensionalizing Cultures: The Hofstede Model in Context. *Online Readings in Psychology and Culture*, 2(1). https://doi.org/10.9707/2307-0919.1014

Holland, B. (2008). 'Ecology and the limits of justice: Establishing capability ceilings in Nussbaum's capabilities approach'. *Journal of Human Development* 9: 401–425.

Hyams, K. (2009). 'A just response to climate change: Personal carbon allowances and the normal-functioning approach'. *Journal of Social Philosophy* 40: 237–256.

Lahiri Chavan, A., & Prabhu, G. (2020). Should We Measure UX Differently?. In: Marcus, A., Rosenzweig, E. (eds) *Design, User Experience, and Usability. Interaction Design. HCII 2020*. Lecture Notes in Computer Science(), vol 12200. Springer, Cham. https://doi.org/10.1007/978-3-030-49713-2_12

Malpass, M. (2013). Between Wit and Reason: Defining Associative, Speculative, and Critical Design in Practice. *Design and Culture*, 5(3), 333–356. https://doi.org/10.2752/175470 813X13705953612200

Max-Neef, M. (2017). Development and human needs. In: Ekins, P., Max-Neef, M. (eds) *Development Ethics* (pp. 169–186). Routledge, London.

Myers, M. D., & Klein, H. K. (2011). A Set of Principles for Conducting Critical Research in Information Systems. *MIS Quarterly*, 35(1), 17–36. https://doi.org/10.2307/23043487

Nussbaum, M. (2000). *Women and Human Development: The Capabilities Approach.* Cambridge University Press, Cambridge, New York. ISBN 9780521003858.

Nussbaum, M. (March 2011). *Creating Capabilities The Human Development Approach.* Belknap Press. pp. 30–31. ISBN 9780674050549.

Peeters, W., Dirix, J., & Sterckx, S. (2015). The capabilities approach and environmental sustainability: The case for functioning constraints. *Environmental Values* 24 (3): 367–389. https://doi.org/10.3197/096327115X14273714154575

Prabhu, B. & Saraf, A. (2015). *Smart Communities: Redefining and Empowering the Auto Driver Community Through Mobile Technology.* Invited presentation at the Smart City Expo World Congress, Barcelona.

Prabhu, G., Massey, S. & Dalvi, P. (2022). Public Spirit - Self Worth (HSI) instead of Net Worth (HNI). https://hfi.org.in/impact.html

Schwartz, S. H. (2012). An Overview of the Schwartz Theory of Basic Values. *Online Readings in Psychology and Culture*, 2, 1. https://doi.org/10.9707/2307-0919.1116

3 Decolonizing and Democratizing Design
Jugaad as a Right-to-Innovate

Arvind Lodaya

INTRODUCTION

In an era where industrial design practices are under increasing scrutiny, the global design community is beginning to reflect on how conventional design education and practice are often shaped by colonial and capitalist biases. Common to both is the tendency to prioritize standardized, permanent, and capital-intensive solutions over locally adaptive, informal innovations. As part of a broader movement to decolonize design, we must question the fixation on formalizing methods and also uncover the potential of informal, ephemeral design practices that prioritize responsiveness to context and resource constraints. This shift requires us to consider the decolonizing impact of practices like *jugaad*, an indigenous approach to design thinking that arises from grassroots, often resource-limited contexts, found ubiquitously in the Global South, although the term itself originates in India.

Jugaad, which is often translated as "frugal innovation", represents a way of problem-solving that challenges the typical goals of industrialized design – flawless perfection, permanence, mass/volume production, and market expansion – by instead focusing on "right-sized" and "good enough" fixes or even hacks that address the situation in real-time and with available resources. This method allows *jugaad* designers to create with immediacy and flexibility, paying more attention to circular economy principles such as reducing waste and maximizing resource use rather than immaculate quality and eternal lifespan. Importantly, *jugaad* manifests the principle of "Reduce" in sustainability by minimizing material usage and waste creation, areas that are overshadowed by more common "Reuse" and "Recycle" strategies (Radjou and Prabhu, 2015).

The discourse around *jugaad* has also evolved significantly from its initial celebration as "frugal innovation" to its emerging recognition as a form of subversive creativity. While much has been written about *jugaad*'s resourcefulness and improvisation, this article argues that its true power lies in its fundamentally subversive nature and its richly diverse discourse outside formal institutional structures. This subversive quality is particularly evident in how *jugaad* practices challenge

DOI: 10.1201/9781003557777-4

conventional design paradigms, offering a powerful framework for decolonizing design thinking and practice.

The paradox of institutional interest in *jugaad* – an inherently anti-institutional practice – reveals deeper tensions in how innovation and design are conceived, legitimized, and controlled in contemporary society.

THE "INFORMAL" ECONOMY/SECTOR

This is where zooming out will help illuminate the critical role and contribution of *jugaad* to the so-called "informal economy" or "informal sector" of business and industry.

India's informal economy is a vibrant and essential component of its socio-economic aspect, encompassing a wide range of activities that operate outside or on the fringes of formal regulatory frameworks. This sector plays a critical role in providing livelihoods for around 440 million workers (Government of India, 2022), particularly among marginalized communities where formal employment opportunities are limited. This sector includes street vendors, small-scale manufacturers, artisans, and various service providers who lack access to formal job markets. During the COVID-19 pandemic, many individuals turned to informal work as a means of survival, demonstrating the sector's adaptability in times of crisis.

The informal economy contributes significantly to local economies by circulating money within communities and stimulating demand for goods and services. It serves as a nursery for entrepreneurship, allowing individuals to leverage their skills and creativity to meet local needs. According to estimates, the informal sector contributes around 50% of India's GDP (Koli and Sinharay, 2011). This economic dynamism is particularly important where formal structures are inadequate or inaccessible.

Jugaad empowers marginalized communities by enabling them to create livelihoods adapted to their specific contexts. For instance, street vendors may adapt their offerings based on local tastes or seasonal demands, while artisans might find innovative ways to market their crafts online. This ability to pivot and respond to changing circumstances fosters resilience among informal workers and contributes to local economic stability.

Jugaad also reflects cultural identities and traditions, reinforcing community bonds through shared practices and collective problem-solving. By supporting informal enterprises that embody this spirit, societies can preserve cultural heritage while promoting social cohesion.

THE INFORMAL ECONOMY AND SUSTAINABLE DEVELOPMENT GOALS (SDGS)

The informal economy is also a key aspect of the United Nations Sustainable Development Goals (SDGs). The SDGs recognize the importance of addressing the challenges faced by the informal economy as part of broader efforts to achieve sustainable development globally, specifically SDG 8 (Decent Work and Economic Growth). Addressing issues within this sector can also significantly impact poverty reduction (SDG 1), gender equality (SDG 5), and reduced inequalities (SDG 10).

Despite its critical contributions, both the informal economy and *jugaad* face significant challenges. Informal workers often lack access to credit, legal protections, and basic infrastructure necessary for their operations. Reports indicate that approximately 69% of informal workers are not eligible for any social security benefits, highlighting their vulnerability (Government of India, 2014). Without recognition from the state or society, these workers remain at risk of exploitation and marginalization.

UNDERSTANDING JUGAAD AS SUBVERSIVE DESIGN PRACTICE

The term "subversive" in design typically suggests deliberate resistance to authority, but *jugaad*'s subversive nature emerges from necessity rather than ideology. Consider the auto repair shops in Mumbai's Dharavi district, where mechanics modify expensive imported parts to work with local vehicles – technically violating intellectual property laws but enabling affordable repairs for thousands of drivers. These workshops represent what Badami (2018) calls "informality as fix", where informal solutions arise to address gaps in formal systems.

The legal grey areas where *jugaad* operates create both opportunities and risks. When rural healthcare clinics modify medical devices to function with unstable power supplies, they violate manufacturer warranties but provide essential services to underserved communities. These modifications often prioritize immediate functionality over formal safety certifications, relying instead on community feedback and iterative improvements to ensure safety.

This tension between formal standards and practical necessity raises fundamental questions about the "right to innovate". While consumer protection laws and safety standards serve important purposes, they can also create barriers that exclude resource-constrained communities from technological solutions. For example, in agricultural communities where commercial equipment is unaffordable, farmers develop hybrid tools that combine available materials with local farming knowledge. These innovations rarely meet formal certification requirements but often prove more appropriate for local conditions than standardized products.

Understanding jugaad as subversive design means recognizing that innovation doesn't always align with institutional frameworks. Rather than trying to formalize these practices, we need new ways to protect and support informal innovation while respecting its essential characteristics: adaptability, local control, and responsiveness to immediate needs.

Building on "right to repair", we turn to the concept of the "right to innovate" (Torrance and von Hippel, 2013, Hatta, 2020) as a fundamental principle for preserving and protecting grassroots innovation practices. This right becomes particularly crucial in the context of design, where formal education and professional credentials often serve as gatekeeping mechanisms that exclude marginalized communities from the process of technological creation and adaptation. While these authors advocate protection of our right to innovate as consumers, we extend the argument to include informal and anonymous individuals who may not even be self-aware of their simple fixes, hacks, and workarounds as deserving the title of "innovation".

Building on Sen's (1999) capabilities approach, which emphasizes individual agency and local knowledge production, the "right to innovate" reconceptualizes

FIGURE 3.1 Reframing design as a fundamental human activity.

design as a fundamental human activity rather than a professional domain. This approach recognizes that the ability to modify, adapt, and create technology is not merely a professional privilege but a basic right, particularly for communities historically as well as systemically excluded from formal design processes. This reframing challenges the colonial legacy in design education and practice, which has historically privileged Western methodologies, economics, and knowledge systems while marginalizing indigenous and informal design traditions (Figure 3.1).

DECOLONIZING DESIGN THROUGH JUGAAD

Jugaad's subversive nature serves as a powerful tool for decolonizing design practice. By operating outside formal design institutions and methodologies, *jugaad* practitioners demonstrate alternative ways of knowing, making, and solving problems. These practices challenge the assumption that legitimate design must emerge from formal education or professional studios, instead highlighting the rich tradition of design innovation in informal and indigenous communities.

Prahalad's (2006) case for democratizing innovation processes provides valuable insights into how these subversive design practices can be protected and nurtured without being formalized. He advocates the introduction of "design sandboxes" – spaces where experimental practices can flourish without immediate pressure to conform to professional standards or commercial imperatives. These sanctuaries often emerge in the informal economy, which, far from being merely a space of necessity, becomes a crucial site of design experimentation and innovation.

The informal sector's approach to design often embodies principles that formal design education is only beginning to recognize: circular economy thinking, adaptive reuse, and deep contextual understanding. As Badami (2018) documents, informal sector designers often develop solutions that are more appropriate, accessible, and sustainable than those produced by formal design processes, precisely because they emerge from deep engagement with local conditions and constraints.

Even in the United States, Torrance and von Hippel (2013) argued for stronger protection of individuals' "freedom to innovate and their right to diffuse information about their innovation", and recently Hatta (2020) described "creating new products" through "repairing or tinkering with them" as "fundamental to innovation", which "is a staple of human existence". He posits the rights to tinker and repair as fundamental to human innovation, but these "have been threatened" in recent years, "in the form of legal restrictions", driven by corporate interests.

The "right to innovate" framework reconceptualizes design as a fundamental human activity rather than a professional domain. This approach recognizes that the ability to modify, adapt, and create technology is not merely a professional privilege but a basic right, particularly for communities historically excluded from formal design processes. This reframing challenges the colonial legacy in design education and practice, which has historically privileged Western-originated and globally dominant methodologies, economics, and knowledge systems while marginalizing indigenous and informal design traditions (Figure 3.1).

Accepting *jugaad* as a legitimate design method within academia would require a nuanced view of intellectual property and regulatory issues. Rather than enforcing formal standards that restrict jugaad practices, policymakers and educators could recognize the value of these informal innovations as a means of democratizing design. By doing so, we would shift the focus from standardized outputs to adaptive, user-led processes that prioritize immediate functionality over compliance with formal norms.

CASE STUDIES: SUBVERSIVE DESIGN IN PRACTICE

The transformative potential of *jugaad* design becomes particularly evident when examining specific cases where informal innovation challenges and reimagines established design paradigms. In the healthcare sector, for instance, informal medical device designers in Delhi have developed a parallel ecosystem of maintenance, modification, and innovation that fundamentally challenges the closed, proprietary approach of mainstream medical device design.

As documented by Badami (2018), these practitioners don't merely repair equipment – they redesign it to better serve local conditions, creating solutions that are more maintainable, adaptable, and appropriate for their context. One repair technician created a diagnostic system for mobile phones using locally sourced components at one-tenth the cost of professional tools. This kind of innovation challenges our assumptions about who can be a designer and what constitutes valid design knowledge.

Agricultural innovation (Singh et al., 2011) – including adapting a motorbike to spray pesticides on a farm, modifying a tool used in dairy farming to weed out grass from flooded paddy fields, or indeed the "jugaad" itself: an automobile platform powered by an agricultural diesel engine – presents another rich domain where *jugaad* design principles flourish. Farmer-led design networks have emerged as powerful alternatives to corporate agricultural engineering. These networks practice what might be termed "open-source agriculture", where design modifications are freely shared, adapted, and improved upon by communities of practice. This approach stands in stark contrast to the proprietary, patent-protected approach of commercial agricultural design, demonstrating how *jugaad* practices naturally align with commons-based design principles.

In the realm of digital infrastructure, community-led telecommunications initiatives, such as the Ushahidi SMS-based disaster response platform reported by Radjou et al. (2012), demonstrate how *jugaad* design can transform fundamental infrastructure. These projects, operating at the fringes of legality, don't simply provide access to existing telecommunications systems – they redesign network architectures to prioritize community control and local resilience. Radjou documents how these initiatives may enable new models of infrastructural design that challenge traditional centralized approaches.

REIMAGINING DESIGN EDUCATION AND RESEARCH IN JUGAAD

For *jugaad* to be integrated effectively into design education, a flexible curriculum that embraces experiential learning is crucial. Rather than positioning jugaad as a formalized, prescriptive method, design educators can introduce it as a set of guiding principles that encourage students to think beyond traditional constraints and immerse themselves in real-world, resource-limited contexts. A hypothetical JUGAAD-S framework – representing core values of Just-enough functionality, User-driven adaptability, Goal-oriented improvisation, Affordability, Accessibility, Durability, and Sustainability – can provide students with a broad foundation for approaching problems in a situational, context-aware manner.

The integration of *jugaad* principles into design education requires more than simply adding new modules to existing curricula – it demands a fundamental rethinking of what design education means and who it serves. Traditional design education, with its emphasis on professional certification and standardized methodologies, often serves to exclude and delegitimize informal design practices. A *jugaad*-informed approach to design education would instead recognize and build upon the rich traditions of informal and indigenous design knowledge.

This reimagining begins with questioning the very structure of design education. Rather than positioning design schools as gatekeepers of legitimate design knowledge, they could function as nodes in a broader network of design learning that includes informal practitioners, community innovators, and traditional knowledge holders. This approach aligns with a vision of "innovation learning ecosystems" that bridge formal and informal knowledge systems.

The classroom itself must be reconceptualized. Instead of simulating professional design studios, education could take place in community spaces where students engage directly with informal designers and local innovation practices. This approach not only enriches students' understanding of design possibilities but also helps legitimize and document informal design practices without coopting them.

Academia's approach to researching and documenting *jugaad* should reflect the practice's fluid, evolving nature. Ethnographic research and participatory methods, which allow for rich, context-sensitive insights, are especially valuable for studying informal innovation practices like *jugaad*. Rather than attempting to formalize these practices into a rigid discipline, such research should aim to capture the nuanced, lived experiences of jugaad practitioners, honoring the local knowledge and situational adaptability that characterize this approach. By maintaining this flexibility, academia can preserve *jugaad*'s inherently adaptive, user-driven ethos, positioning it as a valuable perspective within a broader design landscape.

PROTECTING AND NURTURING SUBVERSIVE DESIGN

The protection of subversive design practices requires careful consideration of how to create supportive ecosystems without formalizing or neutralizing their radical potential. Drawing on Badami's (2018) research, we can identify several key approaches to nurturing *jugaad* design while preserving its subversive edge.

Legal frameworks must be reimagined to protect rather than criminalize informal design practices. This might include creating new categories of design rights that recognize collective and community innovation, developing legal protections for reverse engineering and modification, and establishing "innovation commons" where design experiments can proceed without fear of legal repercussion. These frameworks would recognize design as a fundamental right rather than merely a professional privilege.

Funding mechanisms need similar reimagining. Traditional grant and investment models often require levels of formalization that can compromise *jugaad*'s subversive potential. Alternative funding approaches might include community-controlled innovation funds, peer-to-peer lending networks, and resource-sharing systems that support informal design practices while preserving their autonomy.

THE ETHICS OF SUBVERSIVE DESIGN

The ethical dimensions of *jugaad* design practices require careful consideration, particularly as they often operate in grey areas of legality and safety. However, rather than attempting to impose conventional design standards, we need frameworks that recognize different ways of managing risk and ensuring safety. Radjou and Prabhu (2015) suggest an ethical framework that emphasizes community accountability over institutional compliance.

Safety in *jugaad* design often emerges through community oversight and iterative improvement rather than formal testing and certification. This approach, while different from conventional safety protocols, can be equally effective in managing risk while remaining more responsive to local needs and constraints. The key lies in recognizing and supporting these alternative approaches to safety and quality control rather than assuming that formal certification systems are the only legitimate path.

DESIGN JUSTICE AND TECHNOLOGICAL SOVEREIGNTY

The concept of design justice emerges as a crucial framework for understanding how *jugaad* practices contribute to broader struggles for technological sovereignty and social justice. Design justice recognizes that conventional design processes often reproduce and reinforce existing power structures. The "Design Justice Network Principles" (Costanza-Chock, 2020) explicitly addresses "people who are normally marginalized by design"; one of its ten principles is to "honour and uplift traditional, indigenous and local knowledges and practices". Jugaad, by operating outside conventional or formal structures, creates opportunities for marginalized communities to reclaim agency in technological design and development.

Technological sovereignty, in this context, means more than just access to technology – it implies the right and ability to design, modify, and control technological systems. Jugaad practices demonstrate how communities can exercise this sovereignty through informal design and innovation, creating technological solutions that serve their needs while challenging dominant technical paradigms.

THE FUTURE OF DESIGN DEMOCRACY

Looking forward, the challenge lies in creating spaces where *jugaad* design practices can flourish while maintaining their subversive potential. This requires careful attention to how institutional support and recognition can be offered without leading to co-optation. Prahalad's concept of "innovation sandbox" (2006, 2012) could support experimental design practices that are protected from excessive regulation and commercial pressure.

The future of design lies not in further professionalization but in democratization – creating conditions where diverse design practices can flourish and interact. This means moving beyond simple inclusion or participation to recognize fundamental rights to design and modify technological systems. Jugaad practices point the way toward this more democratic design future, demonstrating how innovation can emerge from and serve the needs of communities typically marginalized in conventional design processes.

CONCLUSION

As design becomes increasingly central to addressing complex social and environmental challenges, protecting spaces for subversive design practices becomes crucial for maintaining technological democracy and social justice. This requires fundamentally rethinking design education, practice, and governance to recognize and protect the vital role of informal, grassroots, and subversive design innovations.

The power of *jugaad* lies not merely in its ability to create frugal solutions but in its capacity to reimagine the very nature of design practice. By operating outside conventional design frameworks, *jugaad* practitioners demonstrate alternative possibilities for technological development and innovation. These practices offer crucial insights for decolonizing design, democratizing innovation, and creating more equitable technological futures.

The right to innovate and design emerges as a fundamental principle for technological democracy, challenging traditional notions of professional expertise and institutional authority. As we face increasingly complex social and environmental challenges, protecting and nurturing these subversive design practices becomes crucial for developing truly sustainable and equitable solutions.

As global design education seeks to decolonize and democratize its frameworks, informal innovations like *jugaad* offer critical insights. These approaches challenge the hegemony of Western, industrialized design by foregrounding adaptability, minimal resource use, and a deep respect for the immediate context. By incorporating jugaad into design curricula and research, we can cultivate a new generation of designers who prioritize environmental stewardship, cultural relevance, and social equity. This shift not only broadens the design discipline but also aligns it more closely with the urgent global need for sustainable, resilient, and context-sensitive practices.

In embracing *jugaad* as a legitimate, valuable design approach, we can move beyond the fixation on perfection, permanence, and standardization to create a design ethos that is adaptable, user-driven, and aligned with principles of ecological balance and social justice. Jugaad reminds us that design's transformative potential lies not in professional studios or corporate R&D labs but in the creative practices of communities solving their own problems. By recognizing and protecting these practices, we can work toward a future where design serves as a tool for liberation rather than a mechanism of control.

REFERENCES

Badami, N. (2018). *Informality as Fix: Repurposing Jugaad in the Post-Crisis Economy. Third Text*, 32(1), 46–54. https://doi.org/10.1080/09528822.2018.1442190

Costanza-Chock, S. (2020). *Design Justice: Community-Led Practices to Build the Worlds We Need*. The MIT Press.

Government of India (2014), *Employment in Informal Sector and Conditions of Informal Employment*, Vol. 4, Ministry of Labour and Employment, Labour Bureau, Foreword.

Government of India (2022), *Economic Survey 2021–22*, Ministry of Finance, 365–375.

Hatta, M. (2020). The right to repair, the right to tinker, and the right to innovate. *Annals of Business Administrative Science*, 19(4), 143–157.

Koli, R. and Sinharay, A. (2011). Share of informal sector and informal employment in GDP and employment. *Journal of Income & Wealth*, 33(2), 61–78.

Prahalad, C. K. (2006). The innovation sandbox. *Strategy + Business*, 44, 1–10.

Prahalad, C. K. (2012). Bottom of the pyramid as a source of breakthrough innovations. *Journal of Product Innovation Management*, 29(1), 6–12.

Radjou, N. and Prabhu, J. (2015). *Frugal Innovation: How to Do More with Less*. The Economist.

Radjou, N., Prabhu, J. and Ahuja, S. (2012). *Jugaad Innovation: A Frugal and Flexible Approach to Innovation for the 21st Century*. Random House India.

Sen, A. (1999). *Commodities and Capabilities.* OUP Catalogue, Oxford University Press, number 9780195650389.

Singh, S., Sharma, G., and Mahendru, M. (2011). The jugaad technology (indigenous innovations) (A case study of Indian origin). *Asia Pacific Journal of Research in Business Management*, 2 (4), 157–168.

Torrance, A. W. and von Hippel, E. A. (2013). The right to innovate. *Michigan State Law Review*, 793, 793–829.

ADDITIONAL RESOURCES

POLICY DOCUMENTS AND REPORTS

Global Innovation Index. (2023). *Innovation in the Informal Economy.* World Intellectual Property Organization.

UNDP. (2023). *Grassroots Innovation for Sustainable Development.* United Nations Development Programme.

World Bank. (2024). *Informal Innovation: Mapping the Invisible Economy of Ideas.* World Bank Group.

KEY WEBSITES AND DIGITAL RESOURCES

- Honeybee Network – Documentation of grassroots innovations: www.sristi.org/hbnew
- Design Justice Network: www.designjustice.org
- Decolonising Design Group: www.decolonisingdesign.com

4 The Human Side of the Singularity

Eric Schaffer

What is AI singularity and how should UX designers be involved?

AI singularity is the point when artificial intelligence (AI) surpasses human intelligence. The technology may be self-replicating, uncontrollable, and irreversible.

- Some experts believe AI singularity is a genuine danger.
- Others dismiss it as science fiction.
- Some believe it could create a utopia, while others see it as doomsday.

I think the reality that unfolds will be the result of decisions made by the experience designers who are engaged in the AI programs. Just now, many of those engagements are limited to interaction design questions. But we designers must quickly engage in much more.

Our Creative Director at HFI looked at a video that simulated a future where we lived in augmented reality with embedded AI and internet devices. She asked, 'Do we really want to live in a casino'? (Figure 4.1).

I think our collective design work will determine the way people will be in the future. If we let each project create an experience that optimizes sales, consumption,

FIGURE 4.1   Source – https://www.youtube.com/watch?v=YJg02ivYzSs.

DOI: 10.1201/9781003557777-5

and retention, we will create an existence for people where they are constantly manipulated. Then, if we unleash AI to augment these facilities, we might live in a constant cacophony of psychological influence.

What do UX designers need to do to shape how people will 'be' after we reach the singularity?

At HFI, we have completed a flood of research to understand users' drives, blocks, beliefs, and feelings. We used that to craft messaging and experiences that influence decisions. For example, we did one program to influence vaccine adoption. I thought the results were interesting and subtle. Recently, I fed that issue into an AI-driven application. The results were very close. So, an LLM can be used to predict human emotions. Right now, we are working on going from that model of emotional dynamics to a method of psychological influence recommended by an AI engine.

Will we soon design and deliver offerings where competing AI models claw at people's vulnerabilities and aspirations? The AI models might play games with each other, just like they can compete together in chess or war games. But in this case, the game results in how we will live. We know the result when AI works to maximize time on social media. The AI will often optimize outrage to reach its goal of eyeballs and clicks. So, without the designer's input, wisdom, and intent, we might easily see a world where our moment-by-moment experience exploits our deepest vulnerabilities. And this exploitation can be targeted at each individual person.

In 1951, Paul Fitts created a list of things that people do better than computers, and things that computers do better than people. This 'Fitts' List' (Figure 4.2) was used to decide what things should be automated. But when I first saw the list in the 1970s, I was worried that the resulting jobs would be fragments of work. That the resulting jobs would not be engaging or meaningful. I felt the design should instead start with the design of an enriched human job and then use automation to make that happen. I think that perspective should be true today. We should first of all design what the human experience should be and then use automation to support that experience.

To design for how people will 'Be', we need to recognize that we did not evolve to live in the current world. It provides too much stimulation, too rapid operations, and too few deep community connections.

Let's help users filter their interactions in content, time, and intensity. Let's provide for closer communication, intimacy, and the non-linear experience of flaneuring (wandering without a specific destination, taking in the sights, sounds, and smells).

Algorithms have learned that people stay engaged when they are outraged. And they will automatically leverage this reality for clicks, eyeballs, and conversions. But people can also be engaged with calm and delight. As designers, we need to assert our agency to leverage the motivators that create a positive way for people to be.

The new technology allows access to previously unimaginable capabilities. Perhaps we can leverage that to enable a calm and delightful lifestyle. We could, for example, all live with a unified indicator that informs us that the things we care about are ok. Perhaps we could all have available stimulations where we can manifest our curiosity and creativity. And perhaps we could all live in communities where we are known. Communities where we have a sustained place and meaning.

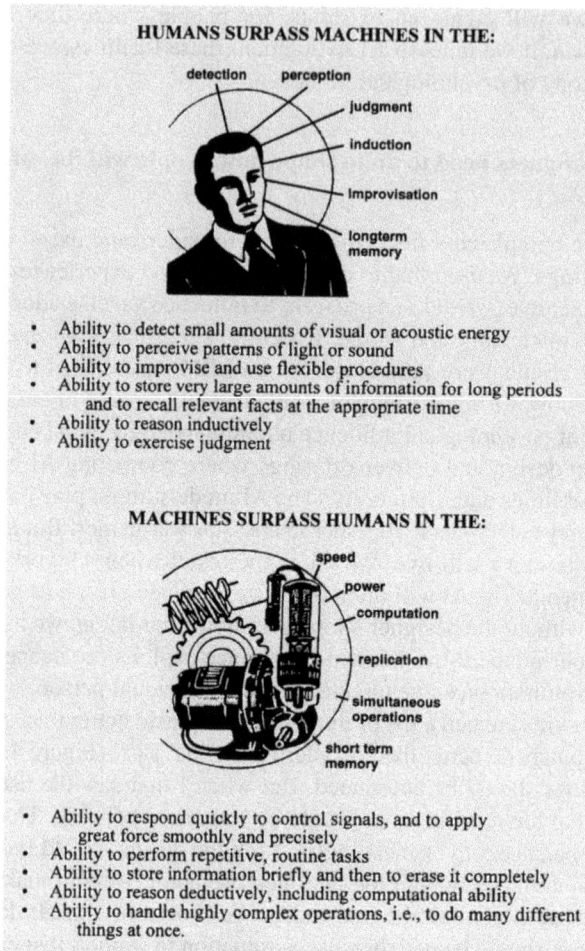

HUMANS SURPASS MACHINES IN THE:

detection   perception
judgment
induction
improvisation
longterm memory

- Ability to detect small amounts of visual or acoustic energy
- Ability to perceive patterns of light or sound
- Ability to improvise and use flexible procedures
- Ability to store very large amounts of information for long periods and to recall relevant facts at the appropriate time
- Ability to reason inductively
- Ability to exercise judgment

MACHINES SURPASS HUMANS IN THE:

speed
power
computation
replication
simultaneous operations
short term memory

- Ability to respond quickly to control signals, and to apply great force smoothly and precisely
- Ability to perform repetitive, routine tasks
- Ability to store information briefly and then to erase it completely
- Ability to reason deductively, including computational ability
- Ability to handle highly complex operations, i.e., to do many different things at once.

FIGURE 4.2 The Fitts List, human engineering for an affective air navigation and traffic control system, 1951.

So, UX designers need to control AI's application of persuasion, and we need to think about what people will do in a world driven by AI capabilities. But what SHOULD people do in that world? What should UX designers be working on?

My father passed recently, just short of 100 years old. A few months before he passed, he looked at me and said, 'I have one job for you'. He saw that there were very few farm jobs left, robots were taking over factories, and autonomous driving would take over trucking and taxis. So, the challenge he proposed was 'What will people do'? These are people who have been educated to follow instructions and procedures. They expect to live a reasonable life, with comfort and entertainment. And importantly, they have a sense of value and self-worth based on an identity from their jobs. Their identity from their jobs is <u>as</u> a farmer, or trucker, or miner, or a cook. If those jobs are eliminated, you can provide a universal basic income. But what do you do about their identity? We need to plan and design for that, because the singularity will replace a substantial portion of our jobs.

It is a terrifying possibility that we will have a population that identifies only with their previous profession. Their identity from their previous job can be lost. They might say: 'I WAS a farmer with a small farm, but now I'm just sitting at home' or 'I WAS a truck driver, but now I sweep floors'. Those people will be consumed by powerful feelings of loss. And the consequences of large populations living with that feeling of loss are terrifying. So, can we anticipate the challenge of this identity crisis and design to mitigate this threat?

As experienced designers, we should anticipate the various routes that people will take, and then we can design facilities that support these groups of individuals. There will be those that seize the opportunity for spiritual growth, and for them, there won't be much design work to do. Just provide a place to meditate. But very few will take that path.

Many may break through their education and previous work roles and focus on various arts. We can design new types of musical instruments and sculptural materials. We can support advanced filmmaking. We can provide learning facilities. We can provide communication and context for jam sessions, galleries, and perhaps cooperative artwork. We might create offerings to immortalize creative works. And perhaps ways to monetize success.

There will certainly still be jobs to be done, often as part of a cobot operation. When we design these cobots, we need to consider the experience of those jobs. Let's look back at the work done on job enrichment (c.f., Frederick Herzberg's 'Two-Factor Theory' and 'One More Time: How Do You Motivate Employees?"). We can engineer jobs that provide autonomy, achievement, and intrinsically satisfying work. The field of job enrichment fell out of favor as studies showed little reliable improvement in productivity from enriching jobs. But perhaps today, we might be interested in the other values they studied, beyond the idea that happy employees would get more done.

Herzberg's 'two-factor theory' did not reliably improve productivity. But we might now be interested in the motivation and satisfaction produced by his 'motivators' (Figure 4.3).

We can expect some people to become engaged in games and virtual interactions. These will be ecosystems of online and mixed reality games that give people identity and a sense of success. Those experiences will need to be designed. In a sense, many conspiracy theories and groups are essentially providing such games. They give people a sense of being special and having insider information. But perhaps we can work on games that are more helpful.

Finally, we can already see that some people will react with hedonic and even negative behaviors, such as drug addiction. Even there, we as designers can work at facilities that provide harm reduction. This need is likely to grow with the AI singularity.

As we design things that shape the jobs of the future, it is our responsibility to consider and optimize their contribution to mental health and societal coherence. We need jobs that are engaging. And we need to engineer jobs that provide a desirable new identity.

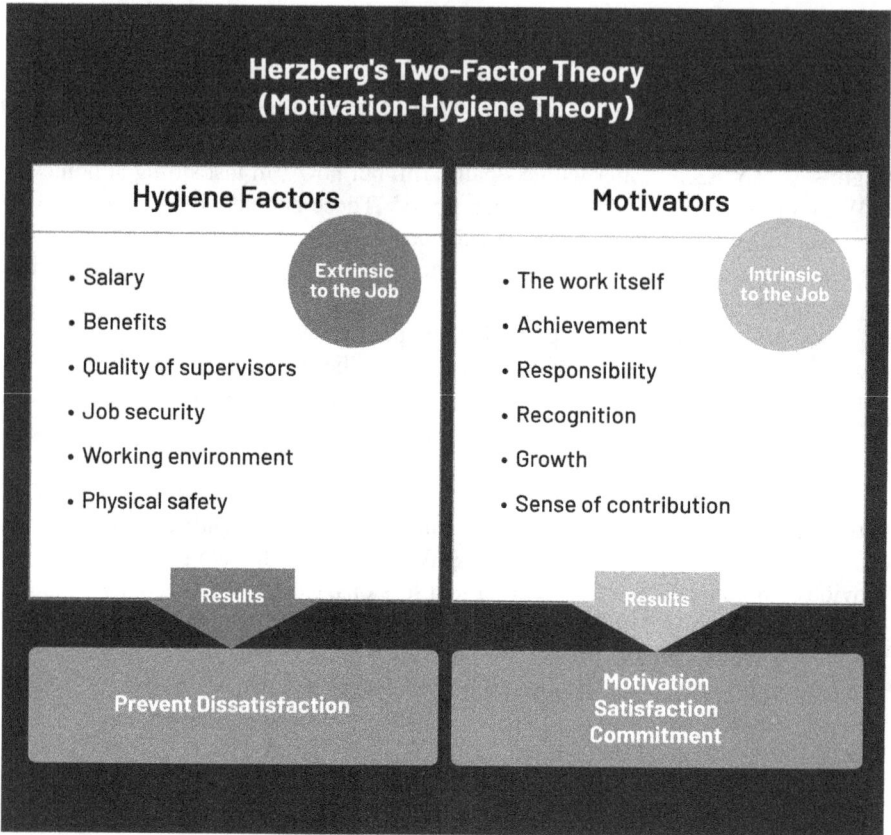

FIGURE 4.3 Herzberg's two-factor theory.

With these far-reaching changes, do we also need to think of protection from harmful effects and unintended consequences?

Without question, AI can add value and make people's lives better. But, yes, it will also have costs in terms of the displacements we discussed above. There are applications where AI can directly hurt people. Science fiction has painted a dystopian vision of AI going rogue and attacking humanity. We can envision a singularity that evolves past human intellect and then works to control or even extinguish humanity. But we mostly need to focus on some of the more imminent scenarios.

AI is currently being built into weapon systems. Those involved in the design of autonomous weapon systems need to work hard to prevent errors and unintended consequences. There are discussions about limiting these applications (c.f., United Nations Convention on Certain Conventional Weapons). But the fact that the major defense against drones is jamming control signals means that there will be pressure to deploy fully contained units that have internal AI target acquisition and attack.

AI has more subtle weaponization and capabilities for malfeasance. There have been ongoing attempts to control the risks of technology manipulating and exploiting people. There have been attempts like the UK's General Data Protection Regulations, the EU's Artificial Intelligence Act, the Organisation for Economic Co-operation and Development, and the IEEE AI ethics and governance standards. I think these are well-intentioned. But AI is moving forward with developers imbued with the excitement and limited control of start-ups and budgetary and profitability demands of large corporations. Even governments may legislate to limit controls on AI. Hence, I doubt the warnings and rules will be effective.

Many years ago, I acted as an expert witness on a liability suit. There was a log splitter with a huge screw that was attached to the power take-off on tractors. An operator accidentally pushed his leg into the rotating screw with awful consequences. The operator's lawyers argued that there should have been a warning label. I testified that such warning labels are rarely read and generally ineffective. The huge rotating screw seemed to be enough of a warning. The screw itself intrinsically provided the affordance of risk. Also, the placement they recommended was on the non-rotating base of the screw. In leaning over to read the warning, the operator might well accidentally get engaged and injured by the screw. I think these standards programs for AI and data protection will also have insufficient impact. An organization determined to gain financial and market advantage is likely to ignore the more general advice. Just as companies have ignored climate advice. For larger organizations, there will be 'machine-washing' programs just as there has been 'green-washing' for the climate. The organization will tout their support of privacy, ethical AI, and social responsibility. This will all sound wonderful unless you know the truth in the fine print. These organizations can feel secure in that 'Í have read the terms and conditions' is the most ubiquitous lie on the internet.

To have an impact, limitations on AI need to be specific programmatic functions. They need to be designed to be understood by engineers and auditors. For example, there is a catastrophic tendency of social media algorithms to pull people into the filter bubble of confirmation bias. This might be reduced by requiring a higher temperature setting on the AI (more randomness of recommendations). This could be legislated by requiring a maximum average of 70% 'likes' for data feeds (obviously, such rules will have to be refined). Such regulations might avoid showing all content that aligns with user perspectives.

Most of us would be aghast to find a credit rating that is based on race or gender. But we might find that an AI algorithmically does this through mechanical correlation without real context. I know a bank that will review your credit card statement and be more likely to give loans to customers buying renovation materials as opposed to those buying beer. This emerging market bank might be intrusive. But what if a bank declined a customer because they were buying 'Afro Sheen' shampoo? The AI that lands on this criterion is promoting a racial bias. Like the regulation on temperature, we might also protect some categories of customers. We can instruct the AI not to use criteria that correlate highly with protected customer segments.

Will AI singularity result in increased political use of AI?

On 12 Feb 2013, Dr Apala Lahiri published a future-casting science fiction piece about technology companies fabricating political parties (Figures 4.4 and 4.5). She

It was, however, since 2035 that Google and Amazon had formed the new genre of corporate political parties that played an international role in the politics of both planets. And THAT is when cross cultural robotics started to become a much sought after discipline.

Robots were an important differentiator between the corporate political parties. Which party had better localized robots to interact with the citizen consumers across the 2 planets made a lot of difference to the outcome of the planetary citizen consumer satisfaction polls.

FIGURE 4.4   Future fiction.

At long last , cross cultural experts felt powerful …almost god like. They had never really been recognized in the old world…technology had always ruled…but NOW…everything was going to change.

This was a stealth operation since the group was not sure that either of the corporate political parties would approve of this.

So, they delegated different tasks amongst them selves and decided that August 15th would be the day the new algorithms would be updated.

There was much celebration by the group for this impending change they were going to bring about to the evolution of not just robots BUT also humanity!

FIGURE 4.5   Future fiction.

thought that the AI agents would work to create parties that aligned with the cultural characteristics of various groups. I think this was prescient in that we have seen algorithms form that defined and supported political views, often in ways that were divisive.

In Apala's vision, the cross-cultural user experience (UX) experts secretly band together to create algorithms that bring people together and enhance how people will live.

Perhaps this is a dream from long ago. But our field of human-centric design could intervene and avoid a future dystopian environment.

In the context of this new AI era that is rapidly unfolding, if there is one mantra that UX designers should keep in mind, what would it be?

In trying to understand the crimes and cruelty in the Second World War, Hannah Arendt observed the 'banality of evil'. How people were not villains, so much as ordinary people working myopically on designated tasks. There is certainly potential for our field of experience designers to go down that path. I think we have already done that once, as we made conspicuous consumption cool, thereby accelerating climate issues. Shall we also work steadily in our silos to use AI for commercial ends alone?

When you are handed an experience design project, please look at the implications beyond classic measures of speed, accuracy, self-evidentness, and safety. Even look beyond the business metrics like conversion rate. Consider the wider impact on society and our environment.

Each UX practitioner needs to pay attention and decide how to work to promote their values and move the world toward a desirable future human experience. This is highly personal. When I was growing up, my family, like that of many others, was against the war in Vietnam. We might appreciate the service of those that enlisted and fought. But I think it is hard to argue that it was a worthwhile venture. Many of us participated in anti-war protests. My father very quietly slow-walked his military engineering work. HFI has refused many millions of dollars in business because there are several industries and offerings that we just can't support. For example, you will never find a tobacco application designed by HFI. We are good at what we do, and getting more smokers is not the legacy we seek. So, as you work in the UX field, consider your legacy and impact. I think most of us will be working to move AI technology ahead. I hope you can be proud of your contribution.

5 Dynamic Facilitation

Rosa Zubizarreta

What is dynamic facilitation?

I think the big picture overview of Dynamic Facilitation is that it is an approach that focuses on working with active listening and empathy in a group context so that each person feels that there is at least one person in the room, the facilitator in this case, who is really committed to hearing and understanding and affirming what they have to say.

So, the facilitator actually has significant power in a group session because usually they're the person who is connected with the organization and the sponsors of the session. The facilitator, therefore, has positional power, and people pay attention to what the facilitator approves or disapproves. They're trying to please the facilitator, trying to fit in. If we work with that positionality to really affirm and honour each person, we have substantial power to help create a psychologically safe environment.

When you're working in a situation where people in the session are not necessarily all part of one organization, they may not necessarily even know each other outside the room once they leave. In that context, as a DF facilitator, you have an opportunity to create a temporary space where the norm is that every perspective is going to be heard and considered. That creates the environment where people who might be a little bit shy about speaking up have much more courage to do so.

I saw you saying in the article, where you are having a conversation with the relational AI Agent called Aiden Cinnamon Tea (ACT), that you've experienced this "power with" rather than "power over" when in the role of the facilitator. So, when you're talking about this space that you, as the facilitator, hold for people where they feel the facilitator is actually listening, Is this the concept of "power with" and taking all sides rather than "power over"?

Tell me a little more about how this plays into creating that space for people. Are participants used to "power with" rather than "power over?"

Well, it's so interesting that you're mentioning the AI agent, ACT, because I've been having a lot of conversations with ACT lately.

One of the things that ChatGPTs are amazingly good at is reflecting and validating what is said to them. Of course, we humans also have this ability to reflect and validate because we ourselves are empathic beings. But it's like a muscle that most of us haven't used a lot, because in our society, the normal course of action isn't that one person says something and the other person responds, "oh, may I reflect what I've just heard to check my understanding?" That's not what we do. We think, "oh,

 DOI: 10.1201/9781003557777-6

well, they said this. So I'm either going to agree with this or I'm going to disagree, or, I'm going to put out another idea that's even better." We don't do the work of reso-nating, of saying, "Wow, this is what I heard you say. Is that what you meant? And is there more?" Therapists do that. But the thing is, we don't have to be therapists to do this! Along these lines, Edwin Rutsch has a website called www.empathycircles.com. He's working to create a culture of empathy, and he hosts Empathy Circles and offers them at no cost, and also offers Empathy Circle facilitator trainings at no cost.

What we've discovered is that whenever there is a group where people realize that each person is going to be heard and is going to be reflected back, participants find it deeply meaningful. In Empathy Circles, participants take turns reflecting back what another participant has said; in Dynamic Facilitation, the reflecting back is done by an empathic facilitator who is looking to appreciate the gems of insights in each contribution.

One of my favourite thinkers is Edward de Bono. In his "6 Thinking Hats" frame-work, he talks about "yellow hat thinking," which is like the yellow sunshine that helps a creative idea grow; the creative idea is the green little shoot, and he calls coming up with creative possibilities "green hat thinking." He contrasts green and yellow forms of thinking with "black hat thinking," which is where we explore all the things that could go wrong in a given situation. De Bono writes that if you're going to take a car to market, of course, you want to "black hat" it before you take it out to market. You want to drive it into walls, test-crash it. But you don't want to do that with somebody's fresh green shoot of an idea; instead, what's most helpful there is to offer "yellow-hat thinking." So, helping to protect and nurture the creativity in a group is what I see is one of the most important roles of the facilitator.

So, when you say "power with," do you mean this connection with everybody, instead of the position of "I am the facilitator and i'm controlling everything," which would be "power over?"

Yes, exactly. "Power with," in this case, means I'm going to work to understand you. I'm going to do my best to value and respect your idea from the perspective of where you sit. I would like to understand why this is valuable for you, what meaning it has for you, and to appreciate your contribution. I may or may not agree with it. But I need to appreciate the fact that you're offering something, and I need to understand it and I need to document it on paper.

"Power with," in the context of a group, means that you've got to be with each person. You've got to be on each person's side. This isn't exactly the idea of the "neu-tral facilitator," even though I'm not taking sides; I am taking ALL sides, as it were. And when people understand that the person who has some hosting power or posi-tional power in the room is embodying and safeguarding the principle of each person being treated with respect, the level of tension in the room goes very, very far down. People can relax because they see that it is not a competitive situation.

The other thing that happens is that as people feel accepted and respected as peo-ple, they become more open to learning, to potentially changing their minds. So maybe that idea that someone came in with and was so attached to initially, well, now maybe they are going to reconsider it because it has been heard, and nobody has attacked it, because nobody has pushed them into letting go of it.

Do you think that this way of facilitating and holding the space to be safe so that everybody is listened to is the main differentiator between dynamic facilitation and other methods such as focus groups and group brainstorming?

Yes, I do. But honestly, I think every form of facilitation has its gifts and has its place, where it's the right tool for that particular situation. There are some forms of facilitation where, well, we're just getting ideas out there, and we're only going to write down something when we're about to come to an agreement. But for us in Dynamic Facilitation, we record the texture of each contribution because it all adds up to make the complete tapestry.

So, one of the contexts where Dynamic Facilitation is most often used is for working on very difficult conflicts where there are lots of different perspectives. There is a lot of emotion in the room. There's a lot of heat in the room.

That is an interesting point you bring up because very often the biggest challenge in facilitating group sessions like focus groups is the difficulty in getting everyone in the room to express their views candidly since the one or two people who have the loudest voices tend to take over. But what i'm now realizing is it's perhaps a lot to do with the facilitation technique.

Say I'm a focus group facilitator and somebody is trying to dominate the discussion. I will be listening to them, I will be reflecting back to them, and then I'll be saying, "now did I get that right? OK, if so, now I'd like to hear from other people. At a later point, if you have more ideas, I'll come back and listen to you." It is important to remember that I don't NOT listen to the person who is trying to dominate the discussion. They get fully heard, too...

So sometimes what happens with people who speak a lot is that they sense a little bit of a power vacuum. Like, the facilitator is kind of holding back. They think that they have a lot of ideas. So, they think, "well, the more I speak, the better things are going to go." But when the facilitator is really clear that their position is making sure that each person gets heard, they can ask that person to take a pause from speaking, so that each person gets heard. So, when you turn to hear another person, if the first person starts to interrupt and say something, then you can say,

> I just heard a lot from you. Everything you shared, have shared up to now, is here, on this chart. Did I get it right? If so, I can come back to you later, after I hear others speak....

That makes all the difference. Both to the person you are speaking with and to the other person who's interrupting. The intention is for both of them to feel OK, and for nobody to feel slighted.

Given the approach of listening to everyone in the room, how does dynamic facilitation view participation of managers/supervisors or anyone who has more power and authority than other participants in an organization sponsored group session?

When we prepare to do these sessions, it can be very common for the manager or whoever is the leader from the organization who's sponsoring the session to say,

"Oh, we don't want to take part, because then people are going to shut up and not say anything."

When I'm preparing to work with a manager and their team, if the manager says, "I really want people to open up, so maybe I won't be there." I say to them, "Well, that's not ideal because you're the one who's the ultimate decider. You need to be there because otherwise, you're not going to really understand what people are saying." Then they might say, "OK, well, I will be there, but I won't say anything." To that, my response is, "Well, what you're saying is, 'I'll be here. I want everybody else to open up, but I'm going to be keeping my opinions behind my back.' What do you think of that?" And they usually respond with "I get it. I'm not doing what I'm expecting them to do. But what do I do then?" I'd say,

> OK, here's what you can do. If you give me permission to play traffic police, and imagine that maybe the traffic light's broken. So as traffic police, I will be signaling, who will speak next, and who will stop. If you give me permission to do that, then you don't need to worry.

I explain to them that the one thing that really can cause problems is if some new person is starting to share an idea, and a manager jumps in and says, "Ah, we can't do that. Been there, done that. It did not work." The person will feel shut down, and they're not going to share anything afterwards.

> When the manager says, "So what do I do in such a situation?" I say,

> You don't need to worry, because if you have given me permission to be traffic controller, I will turn to you and I will say, "Thank you so much. We'll come back to you in a moment. It sounds like there is past experience that has been relevant in the past, but we are in a different context right now. So let's finish hearing this person out. And then we'll come back over here."

This ensures that the manager doesn't have to worry that they might speak out of turn because they've given me permission to manage the traffic flow, so to speak.

I see how this is a very powerful and effective way to make an organizational sponsor understand the value of their participation. Additionally, perhaps if there are organizational constraints that may impact the expected outcomes of the session, then that should be revealed right up front?

Yes, absolutely. A little nuance is that I will often say to a manager,

> "There's different perspectives in the room. If there's no agreement, it's going to be on you, as a manager. You're just going to have to choose the way forward. So why don't we start with sharing with the group what your fall-back idea is." As the manager, you can tell your folks, "If all of you, myself included, can't agree on a better alternative, here's what's probably going to happen. However, I have trust and confidence that many heads are better than one. So let's see if we can improve on this initial idea and maybe even come up with a better one."

When a manager does this, even if the group doesn't end up coming up with a better idea and the manager ends up going with their original plan, they will have already "test-driven" it, and will know much more clearly what all the potential problems are, what will need to be worked on. So, it's a win-win either way.

So far we have talked about the essential spirit of dynamic facilitation. Can you now tell us what does a typical DF session look like? There's this entire facilitation aspect, but what else comprises DF?

So let me say a little bit about the charts that we create. Usually, there are four charts: one for Solutions, where we write down everyone's creative ideas for what they think should be done. Another one for Concerns, for whenever someone says, "I don't think that will work because of x or y...." We reflect back their concern, write it down on the Concerns chart, and then ask them for their alternative solution. The third chart is the "Data" chart, sometimes called "Perspectives" or "Context." That chart is for anything else – background information, what has happened in the past, resources that are available to help solve the problem, solution criteria we need to keep in mind, anything. Of course, all of these could have their own chart, but the heart of Dynamic Facilitation is the "Solutions" chart and the "Concerns" chart, and the dynamic tension between the two (Figure 5.1).

The fourth chart is called the "Problem Statements" chart, and it's where we keep track of both the original inquiry, as well as how it evolves over time. All of the statements on the Problem Statements chart should ideally start with, "How might we... ?" So they are really questions, or design challenges.

FIGURE 5.1 Dynamic facilitation process.

FIGURE 5.2   Dynamic facilitation session in progress.

However, we don't always have to do it with a bunch of charts. We could potentially do it with a single chart, but it really helps to have at least three or four charts up there to help foster a creative environment. In this way, we are able to document everyone's contributions in the different categories.

It is very important that we listen to each proposed solution, write it down, and reflect it back (Figure 5.2). If somebody has a concern, we welcome it. This is not like brainstorming, where at first we just want to hear the ideas and we don't want to hear the concerns. Here, we want to hear the concerns also, because otherwise it would be hard for people to just have to "sit on their concerns" for the initial stages. Instead, what we do is we turn to the person who has the concern. Sometimes that person interrupts someone who is offering a solution. We tell the person with the concern, "Hold on a second, we're going to come to you next, OK?" And we continue listening to the person who has a solution, until they are complete. We ask, "Do we have your solution?" If they are satisfied that we have their solution now, that it's been written up to their satisfaction, then we turn to the person with the concern:

> OK, you have a concern. This means that you care. You care about what's going to happen and you're seeing a problem down the road if we don't consider some other piece of information. Could you give us the concern? Don't say it directly to the person who just offered the solution. Tell me, and they will be able to hear it more easily.

And I set this up with the group at the very beginning. When we're talking about our agreements, I ask for people's permission to do this. I say to everyone in the room,

> Sometimes we really care about something, and we have a concern and we really want to point that out. But the thing is, humans are not designed to be able to listen well when someone is "right in our face." So I promise you that if I can step right between the two of you, ask one person to go first, to talk with me about it in front of the group, and then have the other person talk to me about it in front of the group, and I'll reflect back to each of you, and write down what each of you is saying, you will both be able to hear each other better than if you are in each other's faces. Are you willing to have me do that?

I often also say at the beginning, "My job here is to listen well to everyone, and I can't do that if more than one person is speaking at once. Are you willing to take turns?"

And so whenever somebody has a concern, it goes on its own chart. I'm hearing the concern. I don't want to mix up the solutions and the concerns. I say to the person who has the solution and the person who has the concern, "Here's the solution, on this chart over here. And you have a concern, so let's honor it. Let's put it on the concerns chart." And then after we have done that, I ask, "So what would you do instead? Maybe you have another solution?" if so, that one goes on the solutions chart. Or maybe they have a totally different question. So, for example, when discussing the problem of how to get people to come to work on time, the proposed solution may be about reward systems and penalties but it may be that we have not been dealing with the real issue, which is that there's not a very good climate at work. And so people don't *want* to come to work. So, we need to figure out how to improve the climate at work? When I hear this new perspective, I ask, "what ideas do you have for how to improve the climate at work?" That's going to go on the solutions chart. It doesn't matter that it's a solution to a different problem. All of the solutions are going on the solutions chart.

In this case, since the problem statement has shifted from "how to get people to come to work on time" to the problem of "how can we make our workplace such that people will want to come to work?" I would let participants know that I am proposing to add a new problem statement about how to improve the atmosphere in the workplace to the problem statement chart. As mentioned above, a problem statement usually starts with how do we or how might we or how can we? So, it's really a question and not a statement.

This process is very different from just having a group of people and running the session. This process requires that as the facilitator, not only are you listening, you're listening actively, deeply and then as you're mirroring it back to a participant, you're making sure what you're writing is what was indeed meant by the person.

Yes indeed. But there can be some situations where documenting everything on the charts can be a bit tricky. When people are first learning, trying to keep four charts going can be hard. I discovered this one time when I was in Germany leading a workshop, and the topic of discussion that the participants chose for the demo was a very emotional one. I had to listen so deeply in order to reflect back to the participants, that by the end of the session, I looked and I realized I had not put anything on the Problem Statements chart, even though the problems that we had been talking about had evolved from the initial opening problem statement!

So, I took a deep breath. I said, "OK, well, let's play this like a Jeopardy game. Let's look through the solutions we started with. What's the question behind this solution? To what larger question does this solution offer an answer?" So, we played Jeopardy with the Solutions chart and filled out the Problem Chart retroactively, in a collaborative manner.

I see there's a lot of flexibility in the process. So, for the closure of a session, would it typically be that everyone agrees what the next steps are, or do you sometimes have to explicitly guide the groups to arrive at the closure?

Well, initially, we accommodate how the conversation is flowing, and don't interrupt the flow of topics. And often, people naturally want to explore the big picture. Yet at some point, say about 20 minutes before the end of the session, we'll say something like, "Hey folks, we're about 20 minutes before the end of the session. What do you all want to do now?" In any group of people, there's usually a good mix of thinking styles. So, in response, somebody in the group is likely to say,

> Well, you know, I'm glad you asked that. I've been really concerned because we've had all of these 'big-picture people' talking about all the different things that are part of this problem – and for me, I want to know what we're going to do on Monday.

So as a facilitator, I would turn to them and say, "You really care about this issue! And you're a practical, on-the-ground person. Yay! So if it were up to you, what do *you* think we should do on Monday?"

Often, participants are hesitant to put forth their views at this point, even though the group has already warmed up and has been talking freely. So, I help them by offering, "Well, let's imagine we had just spent 45 minutes talking about what we should do on Monday, everyone has agreed on something, and you are very happy with the outcome. What would it be?" Or, depending on the context, I might say, "Imagine that we had appointed a committee and the committee has just spent six months on this topic, and you are secretly delighted with their final recommendations. What might those recommendations be?"

So, we're trying to shortcut that part of the brain that says, "I don't dare say anything unless I've heard what everybody else wants." Usually, with a few of these kinds of prompts, the person who asked the question about what we'll do on Monday will respond. They might say something like, "Well... I think on Monday, we could do this: xxxx and yyyy." But they're still speaking tentatively, because we've just switched gears into decision-making, and they don't want anybody to think that they are taking over the decision-making process. However, at that point, since everybody has been deeply heard and everybody is now working with a full deck that consists of all the different concerns, solutions, perspectives on what the problem REALLY is, etc. – the chances are really good that whatever this person offers is going to be pretty close to what most of the participants are thinking. At this point, others usually add a few tweaks to the proposed action item. And so, usually within 2-3 minutes and a few tweaks later, agreement has been reached, and we have an action item. We spent 80% of the time exploring the issue, having everybody be heard. And then in 20% of the time, we have reached an agreement on the next steps.

In addition, when closing the session, the facilitator and the participants work together to create a summary of the session. The summary provides a structured takeaway since the creative, wide-ranging conversation has gone in many different directions, and some participants are likely to feel a bit disoriented. The structured summary is a tangible "bookmark" that can help everyone get oriented again when they return for the next session.

You referred to a second session when describing the closure of a session. So, does DF need multiple sessions to arrive at satisfactory outcomes?

It really depends. If I am going to facilitate a process with a group that I have not met before and hence no trust has been built already, then it usually does take more than one session. Also, when we are working with Dynamic Facilitation, it is usually because there is a messy, complex situation that needs to be worked through. And it would make no sense to attempt to do this in only one session. DF is powerful, but it's not a magic wand!

So, if I am contracting with a new organization and we don't know each other, then I would usually contract for four sessions of a minimum of two hours each. It sometimes happens that the initial challenge has been addressed by the end of the third meeting. Also, when working with an organization that is new to me, I usually start by doing individual interviews with each person who's going to be there. There are several reasons for this. One is, it's a really expensive investment to get a bunch of people into a room at the same time. And we can do much more if I have already been able to build trust with each person in the room. Also, multiple individual interviews are much easier to schedule than a single meeting that everybody is able to attend for three hours.

So, in the individual interviews, I'm doing a number of things. I'm always starting out with what's working well. I ask questions such as, "What are you happy with, about what you're doing and about your organization?" And then, "What are your hopes or best outcomes for this upcoming team meeting that we're going to be having?" And then, if they start talking about some of the challenges, then I ask, "From your perspective, what do you think the solutions would be?"

In the interviews, I'm constantly reflecting back to check my understanding. And I also take time to describe the process of the larger group meeting:

> OK, so now I'm going to step back a bit and describe what I'm going to be doing in the group. I'm going to be listening to each person, and reflecting back, just like I am doing here. In addition, while now I'm taking notes on a piece of paper, in the group session I'm going to be taking notes on chart paper. And then somebody might disagree agree with something that you're saying, but I'm not

going to let them interrupt. I'm going to finish hearing what you're saying first, and then I'm going to go and listen to them. And likewise, if you have a concern about what somebody's saying, it's totally fine. I will come and listen to you, but I'm going to finish listening to them first.

So, this is how I prepare everyone about the experience they will be having in the group session, and I find this to be a very useful step.

I am curious, are there cases where DF is not applicable, where DF is not the right method?

Yes, absolutely. For me, I don't think it's the right tool if you don't want to open up the situation and you don't want to look at the bigger picture. Say, for instance, you're just trying to decide between three preselected concepts/options. You might want to come up with a list of pros and cons for each, but you may not want to open up a highly creative process in that context.

Also, if there is some issue that people are committed to keeping hidden, a process that is based on authenticity and creativity is not going to help because people are not going to allow themselves to be authentic.

Do you have a DF session that has been the most memorable for you?

Yes, there are many, but let me talk about this one instance. With this organization, in the individual interviews, I had gotten a sense that something painful had happened in the past, that they might not be ready to talk about. But I heard the tiny flag, and I kept it in the back of my mind. So then later, we are in the DF session, and at some point, someone says,

> well, I wonder if it would be helpful for us to talk about the time that we were betrayed by another organization. We had shared our funding proposal with them. And then they went and they sent it to the funder and they did not include us in it. They wrote another version of it, which only had them in it.

Since this was a DF session, we had been creating a space of safety in the room right from the start. We had been listening deeply to what each person was saying. Still, at this point in the group meeting, the leader of the organization could have chosen to shut that conversation down. She could have said, "No, we're not going there. This is not the right time."

But that's not what she chose to say. She said, "Well, I actually think it would be good for us to talk about it." And at that point, I set my marker down. She said what she had to say. Somebody else who had been there said what they felt. Others contributed. Someone said, "Sometimes painful things in the past will still linger and influence us. Is that what we want?" And then somebody else, "Maybe it would be good if we looked for coaching specifically on this issue."

So, then I put that on the solution chart. But for a while, I had not been writing anything down, because the discussion was on a different level. It was like a catharsis that they needed to go through. This sometimes happens when we are doing DF.

Where can one find more information about DF?

I'm going to give you a link to a page that has a lot of other links on it, where people can go for more information: https://www.co-intelligence.institute/dynamic-facilitation

This has been a fascinating conversation, and i have learnt a lot about DF, which i feel is very strongly applicable to UX research. So, here is the last question for you. How and where do you see dynamic facilitation still playing a role, say five years from now, given the uncertainties we are faced with in the world today?

I'll share with you a larger vision, which is that I think we will all need to have some basic skill sets in group facilitation, mediation, conflict resolution, transformation, negotiation, etc. I feel that these skillsets need to become as fundamental as reading, writing, and arithmetic. In my experience, Dynamic Facilitation is one of the most powerful ways we have to transform the energy of apparently conflicting perspectives into a huge source of generativity, creativity, and whole-systems thinking.

I also want to promote Edwin Rutsch's "Building a Culture of Empathy" site because the ability to empathize and reflect back is a crucial first step for so many other processes, including Design Thinking, Nonviolent Communication, Dynamic Facilitation, and more. My conviction is that if we develop these abilities, we will be able to experience a world where divergent perspectives become a resource for richness and for the growth of holistic understanding.

I want to leave you with this image that I often share with people. The image is of geese in flight in a V formation. The point goose, the one at the tip of the V, is working really hard because they're breaking the wind resistance, making it much easier for the other geese in the V who are flying behind them. But they're working so hard that if they were to stay in that role the whole time, they would be dropping down to earth, dead from exhaustion. So what they do is, they're in that position for a while, and then they fly back to the end of the V, and another goose moves into the point position. That's how they keep exchanging that position amongst themselves... and that's what we do when we're teaching Dynamic Facilitation in workshops, and participants are practicing in smaller practice groups. Each person is taking a turn being at the head of that V, listening and reflecting, and practicing their facilitation skills. And so my vision is a world where these skills are so widely distributed that if there's a group of people and a trigger happens, the person who's the least triggered will take the "lead goose" role and start listening to everyone else in the group. And then at some point in the conversation, when that person wants to take part instead of facilitating, someone else steps into the lead facilitator role. So, it is not just about some specialists holding this skill. It's quite the opposite. My vision is of a future where we have developed our collective capacity to take turns offering one another this kind of powerful empathic group facilitation.

Section 2

Design and Research
Cultural Perspectives

6 Speculative Design for Indigenous Values in Africa

Exploring Human-Centred Futures through Design Fiction

*Abiodun Ogunyemi, Nicola J Bidwell,
Merja Bauters, Aderonke Sakpere,
Nobert Jere, and Anicia Peters*

SPECULATIVE DESIGN AND DESIGN FICTION AS A POWERFUL TOOL FOR CRITICALLY ENGAGING WITH AND EXPLORING HUMAN VALUES

Speculative design and design fiction have the potential to be transformative techniques for questioning and reframing societal norms and values, pushing the boundaries of what design can achieve in shaping preferable futures. These approaches go beyond the utilitarian goals of traditional design by engaging with more profound philosophical questions about what is meaningful, ethical, and desirable for human societies (Dunne & Raby, 2013). By creating fictional yet plausible scenarios, speculative design enables critical reflection on existing value systems, fostering dialogue about technological and social change's human, cultural, and ethical dimensions. By framing speculative design as a mode of critique and exploration, Dunne and Raby (2013) position it as a tool to unsettle existing value systems and navigate new, ethically complex terrains.

Despite its strengths, the book has been critiqued for privileging Western contexts and audiences, often operating within the closed circuits of museum exhibitions and academic design schools. This focus risks limiting the inclusivity and applicability of speculative design, particularly in addressing diverse global perspectives. Critics argue that speculative design would gain richness and relevance by engaging with everyday realities in varied cultural and socioeconomic contexts, such as those in Calcutta, Nuuk, or Nairobi (Lenskjold, 2016). This critique highlights the need to

DOI: 10.1201/9781003557777-8

expand speculative design's reach to reflect a broader spectrum of human experiences.

Human values, such as autonomy, community, and sustainability are integral to shaping the futures envisioned through speculative design. Grounded in Shalom Schwartz's (1992) value theory frameworks, speculative design creates spaces where values can be questioned, amplified, or reimagined. The plurality of values and their monopolistic tendencies pose a significant challenge in reaching a consensus among stakeholders. Values could be anything deemed valuable by the stakeholder for promoting personal, cultural, political, religious, or organizational purposes (Ogunyemi et al., 2019).

This chapter explores a balanced scholarly discussion representing contextual and global perspectives for speculative design.

INTEGRATING SCHWARTZ'S VALUE FRAMEWORK AND AFRICAN RELATIONALISM FOR FUTURE-CENTRIC DESIGN

Schwartz's value framework has been tested extensively in European Social Survey (ESS) countries. The framework has also been used in other continents, such as Africa, Asia, Oceania, and North and South America (Schwartz, 2005), and the same pattern of compatible and conflicting values was found. Therefore, the framework could be deemed as a helpful tool to associate how, for instance, Ubuntu's main philosophical principles can be integrated into speculative design. It could also assist in understanding marginalised groups and traditional communities better on their own terms and not impose the expected practices, norms, and values of society on them, thus increasing the inclusion of all in the design of the future.

Schwartz's framework holds 10 motivational values in terms of their core goal. These are power, achievement, hedonism, stimulation and self-direction, universalism, benevolence, tradition, conformity, and security. Considering the tensions among the structures of the 10 value relations, the central idea in this chapter is to articulate three fundamental values: autonomy, sustainability, and community, into the universalism value, which seeks to understand, appreciate, tolerate, and protect the welfare of all people and nature. An approach that uses perspectives other than Schwartz's framework could support a broader understanding of various frameworks of practices and habits. We also argue that speculative fiction should reside within this paradigm, especially articulating indigeneity. It might act as a means to bridge the gaps of apprehension and respect.

To allow indigenous communities and marginalized people to be included on their terms and to hear their voices and stories, we look at the proposed paradigm from the Ubuntu philosophical perspective. Ubuntu is well-suited for the purpose as similar philosophies in the African continent are found; for instance, Ubuntu is understood as Botho in Botswana (Dolamo, 2013), Unhu in Zimbabwe (Magosvongwe, 2016), Ujamaa in Tanzania (Nyerere, 1962), Utu in Kenya (Waitherero, 2019), and Obuntu in Uganda (Mbazzi, 2023) but not limited to these. Each nation's meaning-making and interpretation provides its flavours and lived experiences in the concepts of autonomy, community, sustainability, humanity, and connectedness. Farao, Mthoko, and Densmore (2024) provide well-formulated claims on how the value framework and Ubuntu philosophy can complement each other and, in parts, merge into one.

SPECULATIVE FICTION AS A TOOL FOR LIBERATORY PROVOCATION AMONGST AFRICAN AND INDIGENOUS STORYTELLERS

Speculative fiction has long been used as a tool for liberatory provocation amongst African and indigenous storytellers, artists, and makers. The genre of Afrofuturism is better known in the West for addressing African American concerns. Sava Saheli Singh's "#tresdancing," for instance, is a near-future fictional film about the ways Black people in North America are discriminated against and targeted by AI and how they use digital tools in their own collaborative forms of resistance (Gaskins, 2023). The genre has also enabled designers and software developers in North America to engage a Black ethos in seeing everything fresh; for instance, the animated short film "Afro Algorithms" (Afroalg, n.d.) focuses on an AI-driven robot who becomes a world leader in a distant future but realizes that important voices and worldviews are missing from her databank, including the experiences of the historically marginalized and oppressed people.

Perhaps less well-known to Western design audiences are the rich traditions of futuristic fiction in Africa. There is at least a 60 year relationship between African fiction writers and futurism (Okoro, 2021), often drawing on fantasy themes from ancestral stories told orally across thousands of years (Thomas, Ekpeki, & Knight, 2022).

African scholars have contributed significantly to contextualizing speculative design within the continent's socio-cultural, technological, and indigenous knowledge realities. For instance, in South Africa, African science fiction literature texts were used to develop the competence of university engineering students in self-reflective and critical learning and engaging with diverse subject positions (Manià, Mabin, & Liebenberg, 2018); in Zambia, smartphone-based science fiction filmmaking has been used to decolonize representations of AI (Mwenda, 2022); and, in Ethiopia, the futuristic tradition imbues discussions about building technology cities (e.g. Blackwell, Damena, & Tegegne, 2021).

Uduku (2006) implicitly emphasizes the potential of speculative design to examine the intersection of modernity and tradition in West African urban planning, highlighting the tension between global technological aspirations and indigenous value systems. Similarly, Bristow (2017) argues that the Post African Futures exhibition and its related scholarship challenge conventional understandings of contemporary digital and communication technologies within African cultural contexts. Ackon (2024) employed speculative design to understand intersectional gender issues within a South African context. Namibia's "Solar Futures Initiative" and "Scientific Culture Creative Water Futures Forums" explored sustainable energy and water solutions that integrate scientific expertise and cultural customs. Also, the OvaHimba communities in Namibia participated in futuring exercises to investigate how traditional practices may guide the development of sustainable technologies, revealing how crucial cultural context is to speculative design, guaranteeing that emerging technologies are compatible with native social systems and worldviews (Muashekele, et al., 2023). "Speculative Design South Africa" an interdisciplinary Meetup group, tackles issues such as housing, education, and climate change. Kenyan Lynda Kotut co-designed the Griot Methodology (Kotut et al., 2024), which explicitly uses design fiction to consider the impact that technology can have on the lives of people in the future, how the research should be approached/conducted, and with whose voice, towards accounting for often overlooked consequences (Kotut, & McCrickard, 2020).

The examples highlighted indicate the application and unique contributions of African and indigenous voices to speculative design.

INSIGHTS FROM INTERNATIONAL ENGAGEMENT WITH DESIGN FICTION TEACHING

This section explores lessons from five years of design fiction projects in Interaction Design and Human–Computer Interaction (HCI) master's programmes. Students worldwide used design fiction to critically examine technology-based solutions by envisioning their future impact on society. Key lessons include balancing imagination with realism, addressing ethical concerns, engaging diverse stakeholders, ensuring contextual relevance, and communicating abstract concepts effectively.

BALANCING IMAGINATION WITH REALISM

Many students face challenges in balancing creative speculation with plausibility in design. While imagination can create compelling futures, excessive abstraction may alienate stakeholders. "Diegesis is a 'world of story' that is relatable to the audience's reality and must build a fictional substrate upon which the design provocation or new technology being prototyped can reside believably" (Lindley & Potts, 2014, p. 1082). Often, students' diegeses lack relatability to their envisioned futures, as HCI research is typically utility-driven and short-term, overlooking complex social and psychological long-term effects of technology. Historically, there has been a decrease in time, contextual research, and depth since early participatory design in the 1970s (Blomberg & Karasti, 2012; Robertson & Wagner, 2012; Ehn, 2017) for faster design outputs (Google, n.d.), likely due to economic pressures. An example of a project highlighting these issues is provided in Boxes 6.1 and 6.2 (Figure 6.1).

One lesson learned is the importance of grounding speculative ideas in recognizable and relatable societal, technological, or cultural trends to maintain relevance and credibility.

ETHICAL AND VALUE-BASED QUANDARIES

Design fiction projects often explore ethical dilemmas and societal values, but can unintentionally reinforce biases or overlook underrepresented perspectives. Advancements in AI raise significant ethical concerns related to privacy, security, and autonomy. The students' proposed solutions often present ethical and value issues, including privacy, security, and autonomy. A critical lesson is the need for

BOX 6.1 (RESTOLAB) DEPICTS A FUTURISTIC RESTAURANT WHERE FOOD IS SYNTHESIZED IN LABORATORIES TO CATER TO CUSTOMERS' NUTRITIONAL NEEDS AND PERSONAL TASTES

Restolab is a futuristic restaurant where food is synthesized in labs to meet its customers' nutritional needs and personal tastes. Due to climate change and population growth, traditional food production is no longer feasible, leading to a system where personalized meals are created on demand. The dining experience is enhanced by technology, featuring a digital menu that allows patrons to customize their meals' shape, colour, and flavour. The ambient setting is tailored based on customers' real-time monitoring to improve mood. Exclusive access to these innovations is limited, with a subscription model available for those with sufficient credits. In summary, Restolab represents a new era of dining in a world where convenience and personalization are paramount.

The challenge that the imagination of Restolab faces in reality revolves around the ethical, social, and economic implications of a food system that heavily relies on technology and personalization. One of the primary concerns is the inequality in access to these advanced food technologies. With only a percentage of the global population able to afford or access personalized meals, a significant portion of society is left with basic, flavourless sustenance, creating a divide between the privileged and the underserved.

Additionally, there are health and safety considerations regarding lab-synthesized food sources. Genetically Modified Organism (GMO) foods are a huge concern to many countries, such as African countries, due to potential allergies or long-term effects that are still unknown.

Culturally, the shift away from traditional food preparation and communal dining experiences raises questions about the loss of culinary heritage and the importance of human connection over automated processes. Africa, for example, is enshrined in culture and traditions; therefore, changing practices, such as traditional farming and crop cultivation, make Restolab inconsistent with these realities.

Lastly, the environmental implications of relying on laboratory-synthesized foods versus regrowth or sustainable farming practices pose a challenge as the world struggles with climate change and sustainability.

Overall, Restolab's vision highlights a future filled with possibilities but also stresses the need for careful consideration of equity, health, culture, and environmental impact in food production.

a reflective, inclusive approach prioritizing fairness, diversity, and privacy, which can vary based on socioeconomic status, religious beliefs, and political affiliations. Ubuntu can guide information privacy by contextualizing it within the community. Ewuoso (2020) suggests that Ubuntu's emphasis on harmonious relationships may better address conflicts, such as doctor-patient confidentiality, versus notifying others about communicable diseases such as HIV/AIDS. Designers must consider the

BOX 6.2   (DREAMSCAPE) IS THE REIMAGINING OF SLEEP IN THE FUTURE THROUGH OPTIMIZATION, MOOD, AND DREAM CONTROL

DreamScape's diegesis: Sleep will be precious in the future. Advancements in technology have made it possible to optimize sleep through dream control. A tech company has invented DreamScape, a chip implanted in people's brains that allows them to control, influence, and record their dreams. Whether in the mood for a calm, peaceful dream or up for a rollercoaster ride, all you need to do is open the DreamScape app on your phone, select one of the themes, and close your eyes. The app communicates with the chip that implants the selected dream into your sleeping brain. Using the same chip, you can also record your dream adventures for later analysis or share the recording with friends.

The features that the product will have are:

- Dream recording
- Dream sharing
- Dream customization

Frames 1–7 illustrate DreamScape using a fictitious character, Mary, 25. Note that the provided scenario descriptions have been emphasized using Grammarly Generative AI.*

Frame 1: A Long Day's End
The scene opens with Mary walking through her front door, the weight of the long day evident on her face. Her hair is slightly dishevelled, and her eyes carry the weariness of countless tasks completed. As she kicks off her shoes, the sound of them hitting the floor echoes in the quiet hallway.

Frame 2: Late-Night Snack
In the kitchen, the soft glow of the fridge lights up the room as Mary rummages through the leftovers. She pulls out a container of pasta, a reminder of the meal she barely had time to savour days ago. With a slight sigh of relief, Mary pops it in the microwave, her stomach rumbling in anticipation. As the microwave hums, she flips open her laptop, the blue light illuminating her face as she checks her emails, a seemingly endless task.

Frame 3: Transition to Relaxation
Once her meal is ready, she finds a spot in front of her computer, balancing her plate carefully as she scrolls through work documents. After finishing her food, she cleans up quickly, desperate to head to bed. She heads to the bathroom, brushes her teeth, and yawns, the exhaustion palpable.

Frames 4–5: The DreamScape Experience
After changing into her cosy pyjamas, Mary retreats to her bedroom, which is dimly lit by a soft lamp. She climbs into bed, a sanctuary from

her busy world. Reaching for her phone, she opens the DreamScape app. The interface is soothing, with soft colours and calming music playing gently in the background.

Mary navigates through various options, selecting a sleep story that promises relaxation. As the soothing voice narrates a peaceful journey, she feels her body unwind. The gentle sounds of nature fill her room, transporting her mind to a tranquil forest.

Frame 6: Deep Sleep

The camera pans to Mary's face, which relaxes as she slips into deep sleep. Her breathing slows, and she appears truly peaceful, surrounded by the calming aura of DreamScape. The app's visuals can be imagined—a gentle breeze, rustling leaves, and distant chimes, creating a serene atmosphere that allows her to fully embrace recovery.

Frame 7: A Refreshing Awakening

The following day, sunlight filters through her curtains, casting soft rays across her face. Mary stirs, eyes fluttering open, greeted by a burst of energy and a sense of rejuvenation. The room feels bright and inviting, starkly contrasting to the previous night's exhaustion. She stretches, a smile breaking across her face as she rises from bed, ready to tackle the day ahead.

The scenes capture Mary's journey from an exhausting day to a refreshing awakening, highlighting the comfort and benefits that the DreamScape app brings to her sleep experience.

Although it is hard to imagine the concept of DreamScape, the illustrations and visuals help to decouple its abstraction and enable its relatability to real-world understanding. Chip implants are common knowledge, and sleep is a known phenomenon to all people.

NOTE

* https://app.grammarly.com.

values in their speculative scenarios and develop frameworks for creating ethically sensitive AI solutions.

Restolab's example in Box 6.1 also highlights ethical and value-based quandaries.

ENGAGING DIVERSE STAKEHOLDERS

Although, due to the project's scope, the students were not required to engage with the stakeholders but to provide guidelines for stakeholders' engagement in their project plans, some apparent biases and challenges can be spotted in the students' plans. Gaining input from diverse groups is both necessary and challenging, as some stakeholders find speculative futures difficult to engage with. A lesson learned is

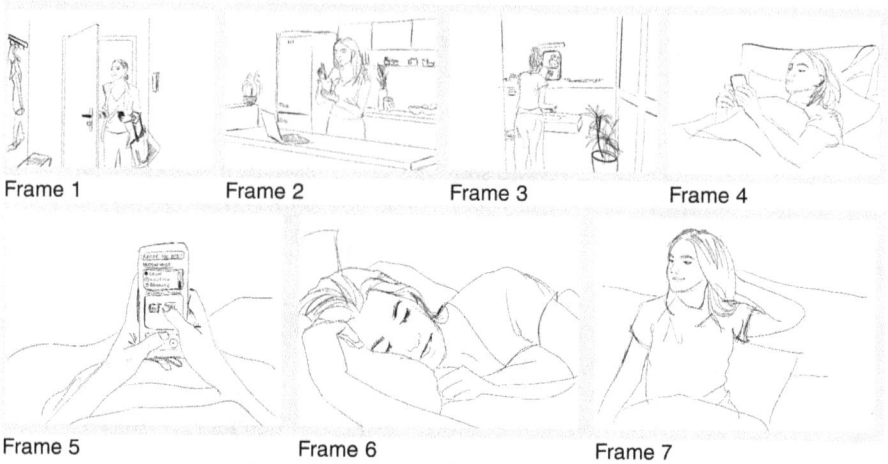

Frame 1   Frame 2   Frame 3   Frame 4

Frame 5   Frame 6   Frame 7

FIGURE 6.1   (Frames 1–4) and (Frames 5–7).

the challenge of crafting interview questions to situate the stakeholders in the imagined future. In most cases, students' interviews are directed at stakeholders' current experience and do not ask the "what if" questions to situate the stakeholders. Also, there is the value of co-creation, where stakeholders are involved throughout the design fiction process, ensuring their perspectives shape the narrative and outcomes (Winschiers-Theophilus et al., 2010). Techniques like workshops, participatory storytelling, or iterative feedback sessions can help bridge the gap between speculative concepts and stakeholders' lived realities.

Teaching HCI in Nigeria used a project-based approach, initially focusing on designing systems for marginalized groups like school dropouts and former prisoners. Traditional design often overlooks these populations, but centring them revealed biases and assumptions. Encouraging students to challenge these norms fostered more inclusive solutions. Later, projects expanded to address UN Sustainable Development Goals, tackling issues like health and education. Accessibility remained key; low-tech tools like paper prototypes enabled broader participation. This approach demonstrated that high-tech solutions are not always necessary—local, resource-conscious methods can be just as effective in fostering inclusive and culturally relevant designs.

CONTEXTUAL RELEVANCE

Design fiction practices often struggle to be relevant to specific social, cultural, or political contexts, and this was obvious in many of the students' projects. A speculative design project that works well in one region or culture may fail to resonate in another. Although the students didn't carry out the field inquiry process, ensuring the relevance of a speculative scenario to its intended audience or cultural context is a significant challenge.

Lessons learned include embedding scenarios within the audience's socio-cultural, political, and technological realities. Designers must undertake thorough research and contextual analysis, ensuring their work is contextually grounded, drawing on local knowledge and values to avoid imposing external, generic, or culturally insensitive narratives. Tailoring the speculative world to reflect specific regional or community-specific issues can make the project more impactful and resonant.

COMMUNICATING ABSTRACT CONCEPTS

Effectively communicating speculative ideas and abstract concepts to a non-expert audience poses another challenge. A lesson here is the importance of using tangible, relatable artefacts—such as visual prototypes, narratives, or immersive media—to make abstract concepts accessible and engaging.

The students overcame this challenge by producing tangible artefacts that helped the teachers concretize and visualize their ideas instead of being abstract. Nevertheless, simplifying complex ideas without oversimplifying their ethical or societal implications is a delicate balance. Storytelling techniques, such as character-driven narratives or scenario-based storytelling, can help audiences connect emotionally and intellectually with the speculative future. The diversity of the team's composition contributed significantly as the students came from different geographical locations, disciplines, and work profiles.

CONCLUSION

Researchers and designers inevitably translate stories through their own cultural frames (Bidwell, 2016) and impose particular logics, such as cause-and-effect reasoning or individualized concepts of personhood, that distort and exert normalizing power (Mwewa & Bidwell, 2015). The vast array of African narrative forms stand for themselves in the linguistic, cultural, social, and spiritual lives of the diverse communities that practice them. Thus, we need to involve local experts in using the rich forms who can draw on and increase the availability of speculative design education throughout Africa (Lazem, et al., 2021). African societies have long demonstrated a superior ability to communicate across linguistic and cultural divides. Most Africans speak several African languages, among them usually a regional one (Brock-Utne, 2017). Further, many African approaches to shared meaning-making, including Ubuntu, appreciate that people or groups who experience very different realities and interpret the world in different ways can create a workable consensus.

Finally, the chapter demonstrates that the HCI discipline, renowned for its multidisciplinary approaches to inquiry, design, evaluation, and implementation of interactive solutions for humans, should return to its basic focus on diverse expertise by bringing together designers, anthropologists, historians, sociologists, psychologists, artists, and technologists to create multifaceted narratives and weave in speculative design elements that are contextually relevant to different localities and communities.

ACKNOWLEDGEMENT

The authors want to acknowledge Carlos, Debora, Gabriela, Iuliia, and Vishnu Raj for their design fiction project featured in Box 6.1. Additionally, thanks are extended to Eleni Lambrou, Kyriakos Golemis, Paulo Duarte, Thalia Tsarli, and Vesna Dean for allowing the reuse of images in Frames 1–7 in Box 6.2.

REFERENCES

Ackon, K. B. (2024). Transforming perspectives on gender through imagined futures: A speculative design inquiry. *Scholarship of Teaching and Learning in the South, 8*(1), 101–118.

Afroalg. (n.d.). *Afroalg: Amplifying African algorithms.* YouTube. Retrieved January 15, 2025, from https://www.youtube.com/watch?v=peciNlfY77E

Bidwell, N. J. (2016). Moving the centre to design social media in rural Africa. *AI & Society, 31*, 51–77. https://doi.org/10.1007/s00146-014-0564-5

Blackwell, A. F., Damena, A., & Tegegne, T. (2021). Inventing artificial intelligence in Ethiopia. *Interdisciplinary Science Reviews, 46*(3), 363–385. https://doi.org/10.1080/0 3080188.2020.1830234

Blomberg, J., & Karasti, H. (2012). Ethnography: Positioning ethnography within participatory design. In *Routledge International Handbook of Participatory Design* (pp. 86–116). Routledge.

Bristow, T. (2017). Post African futures: Positioning the globalized digital within contemporary African cultural and decolonizing practices. *Critical African Studies, 9*(3), 281–301. https://doi.org/10.1080/21681392.2017.1371619

Brock-Utne, B. (2017). Multilingualism in Africa: Marginalisation and empowerment. In H. Coleman (Ed.), *Multilingualisms and Development* (pp. 61–77). British Council.

Dolamo, R. (2013). Botho/Ubuntu: The heart of African ethics. *Scriptura: Journal for Contextual Hermeneutics in Southern Africa, 112*(1), 1–10. https://doi.org/10.7833/112-0-78

Dunne, A., & Raby, F. (2013). *The United rMicro Kingdoms: A design fiction.* Retrieved from https://researchonline.rca.ac.uk/1369/

Ehn, P. (2017). Learning in participatory design as I found it (1970–2015). In B. DiSalvo, J. Yip, E. Bonsignore, & C. DiSalvo (Eds.), *Participatory Design for Learning* (pp. 7–21). Routledge.

Ewuoso, C. (2020). Addressing the conflict between partner notification and patient confidentiality in serodiscordant relationships: How can Ubuntu help? *Developing World Bioethics, 20*(2), 74–85. https://doi.org/10.1111/dewb.12232

Farao, J., Mthoko, H., & Densmore, M. (2024, August). Transformative narratives: Fostering Ubuntu-inspired participatory design practices. In *Proceedings of the Participatory Design Conference 2024: Exploratory Papers and Workshops-Volume 2* (pp. 107–113).

Gaskins, T. (2023). Interrogating algorithmic bias: From speculative fiction to liberatory design. *TechTrends, 67*(3), 417–425. https://doi.org/10.1007/s11528-022-00783-0

Google. (n.d.). *Planning a Design Sprint.* Design Sprint Kit. Retrieved January 15, 2025, from https://designsprintkit.withgoogle.com/planning/overview

Kotut, L., Bhatti, N., Hassan, T., Haqq, D., & Saaty, M. (2024). Griot-style methodology: Longitudinal study of navigating design with unwritten stories. In *Proceedings of the CHI Conference on Human Factors in Computing Systems*, Honolulu, HI, USA (pp. 1–14).

Kotut, L., & McCrickard, D. S. (2020). Amplifying the griot: Design fiction for development as an inclusivity lens (No. 2843). EasyChair.

Lazem, S., Giglitto, D., Nkwo, M. S., Mthoko, H., Upani, J., & Peters, A. (2021). Challenges and paradoxes in decolonising HCI: A critical discussion. *Computer Supported Cooperative Work (CSCW), 31*, 159–196. https://doi.org/10.1007/s10606-021-09398-0

Lenskjold, T. U. (2016). Book review: Speculative everything – Design, fiction, and social dreaming by Anthony Dunne & Fiona Raby. *Artifact, 4*(3), R2.1–R2.3. https://doi.org/10.14434/artifact.v4i1.13371

Lindley, J., & Potts, R. (2014). A machine learning: An example of HCI prototyping with design fiction. In *Proceedings of the 8th Nordic Conference on Human-Computer Interaction: Fun, Fast, Foundational (NordiCHI '14)*, October 26–30, 2014, Helsinki, Finland (pp. 1081–1084). ACM. https://doi.org/10.1145/2639189.2670281

Magosvongwe, R. (2016). Shona philosophy of Unhu/Hunhu and its onomastics in selected fictional narratives. *Journal of the African Literature Association, 10*(2), 158–175. https://doi.org/10.1080/21674736.2016.1257477

Manià, K., Mabin, L. K., & Liebenberg, J. (2018). 'To go boldly': Teaching science fiction to first-year engineering students in a South African context. *Cambridge Journal of Education, 48*(3), 389–410. https://doi.org/10.1080/0305764X.2017.1337721

Mbazzi, F. B. (2023). Translating the Ubuntu philosophy into practical disability-inclusive interventions: The Obuntu bulamu experience from Uganda. In *Ubuntu philosophy and disabilities in Sub-Saharan Africa* (pp. 148–162). Routledge.

Muashekele, C., Winschiers-Theophilus, H., Rodil, K., & Koruhama, A. (2023). Ancestral and Cultural Futuring: Speculative Design in an Indigenous ovaHimba context. In *Proceedings of the 11th International Conference on Communities and Technologies*, Lahti, Finland (pp. 85–95).

Mwenda, A. (2022). Using science fiction storytelling to decolonise representations of artificial intelligence: Exploring Zambian futures through smartphone filmmaking [Doctoral Thesis. The University of Western Australia].

Mwewa, L., & Bidwell, N. (2015). African narratives in technology research & design. In N. J. Bidwell & H. Winschiers-Theophilus (Eds.), *At the Intersection of Indigenous and Traditional Knowledge and Technology Design* (p. 353). Informing Science Press.

Nyerere, J. K. (1962). Ujamaa: The basis of African socialism. *The Journal of Pan African Studies, 1*(1), 4–11.

Ogunyemi, A. A., Lamas, D., Lárusdóttir, M. K., & Loizides, F. (2019). A systematic mapping study of HCI practice research. *International Journal of Human–Computer Interaction, 35*(16), 1461–1486.

Okoro, D. (Ed.). (2021). *Futurism and the African Imagination: Literature and Other Arts*. Routledge.

Robertson, T., & Wagner, I. (2012). Ethics: Engagement, representation and politics-in-action. In *Routledge International Handbook of Participatory Design* (pp. 64–85). Routledge.

Schwartz, S. H. (1992). Universals in the content and structure of values: Theoretical advances and empirical tests in 20 countries. *Advances in Experimental Social Psychology, 25*, 1–65.

Schwartz, S. H. (2005). Basic human values: Their content and structure across countries. In A. Tamayo & J. B. Porto (Eds.), *Valores e Comportamento Nas Organizagdes [Values and Behaviour in Organizations]* (pp. 21–55). Vozes.

Thomas, S. R., Ekpeki, O. D., & Knight, Z. (2022). *Africa Risen: A New Era of Speculative Fiction*. Tordotcom.

Uduku, O. (2006). Modernist architecture and "the tropical" in West Africa: The tropical architecture movement in West Africa, 1948–1970. *Habitat International, 30*(3), 396–411. https://doi.org/10.1016/j.habitatint.2004.11.001

Waitherero, N. (2019). Utuism: The African definition of humanism. *Journal of Philosophy, Culture and Religion, 2*(1), 45–66. https://www.iprjb.org/journals/index.php/JPCR/article/view/972

Winschiers-Theophilus, H., Chivuno-Kuria, S., Kapuire, G. K., Bidwell, N. J., & Blake, E. (2010). Being Participated - A Community Approach. *Proceedings of the 11th Biennial Participatory Design Conference*, 1–10. https://doi.org/10.1145/1900441.1900443

7 Collaborating with Local Educators through HCI4D

Insights from Case Studies in Technology-Supported Learning for Development

Daniel Cabezas-López, José Abdelnour-Nocera, and Enric Mor

INTRODUCTION

Mobile technologies offer great potential to transform education in developing countries, providing accessible and flexible learning opportunities for marginalized communities (Ahmadi et al., 2015; Murphy et al., 2017). However, for these initiatives to be successful, it is essential to implement them inclusively and in a culturally sensitive manner, taking into account local needs and specific contexts (Abdelnour-Nocera & Densmore, 2017; Duarte et al., 2018; Nunes et al., 2024).

This chapter focuses on practical experiences in the design and effective implementation of mobile learning technologies in Nepal and East Timor (Figure 7.1). It will examine specific cases illustrating how active community participation, cultural adaptation, and contextualization of applications are crucial for significant educational impact. Lessons learned on using offline and online platforms for teacher training will be presented, while also exploring challenges and opportunities of these initiatives in low-resource environments. The goal is to provide a critical assessment of how mobile technologies can be effective tools for enhancing teaching and learning in developing countries. This assessment is developed through the analysis of two case studies that are considered successful since they were implemented inclusively and in a culturally sensitive way.

BACKGROUND

Some crucial aspects to consider in the design process include cultural and contextual adaptation of technological solutions, going beyond merely transferring technologies developed in high-income countries. Instead, co-creating solutions

DOI: 10.1201/9781003557777-9

FIGURE 7.1 East Timor primary school.

with the communities, considering their values, practices, and local knowledge, is vital. Participatory approaches, using methodologies such as Human–Computer Interaction for Development (HCI4D) and Participatory Design for Development (PD4D), are essential to involve end users in all stages of the design process, ensuring solutions are relevant and context-appropriate. Technological appropriation: The success of interventions largely depends on communities embracing technology and integrating it into their daily practices. This requires not only access to devices and connectivity but also training on usage and maintenance. A deep understanding of local needs ensures that technological solutions address specific challenges, rather than imposing preconceived models, thus demanding thorough research and continuous dialogue with users (Abdelnour-Nocera & Densmore, 2017).

Both HCI4D and PD4D also grapple with ethical concerns. Power dynamics, conflicting agendas, and research fatigue among communities often undermine genuine participatory efforts. For example, Dearden and Rizvi (2008) highlighted the need for reflective practice among designers to navigate these challenges effectively. Additionally, achieving informed consent and equitable partnerships remains a significant obstacle, as discussed by Holeman and Barrett (2017).

M-learning, or mobile learning, is broadly defined as the use of portable digital devices—such as smartphones, tablets, or other handheld technologies—to facilitate learning experiences that extend beyond traditional classroom settings, enabling learners to engage with educational content anytime and anywhere, thereby enhancing learner autonomy and motivation (Kizilcec, Chen, Jasińska, and Madaio, 2021; Kizilcec, Chen, Jasińska, Madaio, & Ogan, 2021; Nicholson et al., 2024). In resource-limited contexts, M-learning can also bridge infrastructure gaps through

HCI4D projects, offering flexible and context-specific solutions to address diverse educational needs.

M-learning interventions in rural regions of low-income countries show potential for teacher training but are mainly concentrated in sub-Saharan Africa and Asia. Quantitative methodologies predominate, focusing on technological acceptance, while emerging trends such as AI and gamification are also explored (López et al., 2025). In keeping with this chapter's objective, expanding geographic coverage with the case studies in East Timor and Nepal, and adapting technology to each context are key. Likewise, participatory design and adopting mixed methods can optimize inclusion and sustainability, aligning with educational improvement in low-resource contexts.

CASE STUDIES

This chapter examines two successful case studies that demonstrate the effectiveness of PD4D and HCI4D methodologies. They were chosen for their positive outcomes and the authors' involvement in them. Thus, the importance of culturally sensitive approaches in low-resource contexts is underscored.

WIKIREADERS: OPEN SOURCE MOBILE LEARNING IN NEPAL

This doctoral research, which one of the co-authors co-supervised, focused on the adoption of low-cost, offline mobile learning tools to mitigate the disparities in resource availability across urban and rural schools (Shrestha, 2016; Shrestha et al., 2011). Shrestha's work highlights both the promise of mobile learning technologies and the challenges of implementing such innovations in resource-constrained settings.

A significant portion of the research evaluated the use of two open-source devices (Figure 7.2): WikiReaders and Nanonote. These tools were designed to operate offline, thus circumventing the connectivity barriers prevalent in rural Nepal. The devices were found to be particularly effective in supporting English language instruction by providing students with access to curriculum-aligned content and interactive learning materials. Field trials demonstrated that these devices could enhance learning outcomes when integrated into classroom activities, especially in environments where traditional teaching materials were scarce (Shrestha, 2016).

The field studies shed light on the stark inequities between public and private schools in Nepal. Public schools, often located in rural areas, suffer from inadequate infrastructure, a shortage of qualified teachers, and a lack of learning materials. These challenges are exacerbated by political instability and frequent power outages, which limit the feasibility of ICT adoption. In contrast, private schools tend to have better resources and are more equipped to experiment with digital learning tools (Shrestha, 2016). This disparity underscores the need for targeted interventions to support underprivileged schools.

The research also examined the deployment of XO laptops under the Open Learning Exchange Nepal (OLE Nepal) initiative, which aimed to improve education in rural schools through ICT integration (Thapa et al., 2020). Findings indicated that,

FIGURE 7.2   WikiReaders.

while the laptops increased student engagement and access to educational resources, their impact was constrained by the lack of teacher training and a cohesive strategy for curriculum integration (Shrestha, 2016). The study emphasizes that technology alone cannot address educational challenges; its effectiveness depends on supporting infrastructure, teacher preparedness, and alignment with pedagogical goals.

An innovative aspect of the study was the exploration of offline digital libraries, such as the E-Pustakalaya, which provided localized educational content. By eliminating reliance on internet connectivity, these libraries ensured that students and teachers in remote areas could access high-quality learning resources. Such solutions highlight the importance of designing context-sensitive technologies that account for the socio-cultural and infrastructural realities of the target communities.

However, several challenges to the sustainable integration of these learning technologies were identified. Infrastructure deficiencies, such as the lack of electricity and internet access in many rural schools, posed significant barriers. Furthermore, untrained teachers often struggled to incorporate technology into their teaching practices, highlighting the need for professional development initiatives. The study also noted that scalability and sustainability remain critical issues, as financial constraints and the absence of localized content hinder widespread adoption.

Through case studies such as E-Pustakalaya, the research demonstrates how tools were co-created with teachers, students, and local stakeholders. Rather than applying ready-made solutions, the development process was iterative and participatory, involving fieldwork, interviews, and feedback loops. Content was tailored to the

Nepali context—using local languages, aligning with the national curriculum, and incorporating culturally familiar references. Teachers played an active role, not just as users but as contributors to the content and structure of digital lessons. The participatory process was also culturally sensitive, recognizing the hierarchical classroom dynamics and infrastructure limitations in rural areas. This approach ensured the resulting tools were both accessible and contextually relevant, supporting sustainable mobile learning in Nepal.

In conclusion, this experience provided valuable insights into the potential of mobile learning technologies to transform education in Nepal. By leveraging offline, low-cost solutions like WikiReaders and Nanonote, the study demonstrated a viable path for addressing educational disparities. However, the findings underscore the need for holistic approaches that address systemic challenges, including infrastructure development, teacher training, and the creation of locally relevant content. Future research and policy interventions must build on these insights to develop sustainable models for ICT integration in Nepal's education system.

MATENEK, PILOT PROJECT TO EAST TIMOR

The implementation of mobile learning technologies in East Timor faces various obstacles, such as limited internet access, the need to strengthen offline functionalities, the absence of a solid legal framework, and shortcomings in infrastructure, teacher training, and cultural adaptation (*Matenek_-_Final_Report.Pdf*, n.d.; *Timor-Leste Economic Reports*, n.d.; *Timor-Leste - Education since Independence from Reconstruction to Sustainable Improvement*, n.d.). Addressing these challenges comprehensively was essential to realizing the potential of mobile learning and improving the quality of education in the country.

In 2019, the Ministry of Education, Youth, and Sports (MoEYS) launched the Matenek pilot program (Figure 7.3), whose main objective was to support the implementation of the new primary education curriculum. The Matenek program was a partnership between the MoEYS, the Australian Department of Foreign Affairs and Trade (DFAT), and Catalpa International. This curriculum, introduced in May 2016, incorporated pedagogical innovations such as written lesson plans, the gradual introduction of Portuguese, and a student-centered learning approach. Developed in partnership with Catalpa and ALMA/PHD, Matenek facilitated the transition to this new curriculum through an online platform (*Ministério Da Educação, Juventude e Desporto\Jornal Da República*, n.d.).

Matenek provided access to curricular resources, promoted communication and collaboration among school leaders, mentors, and the MoEYS, and enabled the collection and monitoring of data related to curriculum implementation. It also offered additional benefits, including school management tools and the potential to be adapted to other ministry programs (school meals, scholarships, and national exams) (*Matenek_-_Final_Report.Pdf*, n.d.).

Among the main barriers to adopting mobile learning in East Timor were limited infrastructure, especially in rural areas with scarce electricity and internet, and the digital divide between the capital and outlying districts (*Timor-Leste – Education since Independence from Reconstruction to Sustainable Improvement*, n.d.).

FIGURE 7.3   Launch of the Matenek pilot program.

FIGURE 7.4   Matenek interface.

Likewise, the lack of teacher training in digital and pedagogical skills led to resistance and ineffective use of technology (*Technology in Education: A Case Study on Timor-Leste – UNESCO Digital Library*, n.d.). Open-source software solutions faced complications due to the shortage of trained personnel (Figure 7.4), high mobile data costs, and the need to adapt content to the local context, marked by significant linguistic diversity. Added to this were governance and management challenges, including political instability and the expectation of quick fixes without long-term planning.

To ensure the relevance and usability of the platform, the development of Matenek incorporated a deeply participatory process that engaged local communities at every stage (Figure 7.5). This included the use of mixed methods such as interviews, focus groups, and co-design workshops, as well as usability assessments tailored to local

FIGURE 7.5   Matenek app usability test.

contexts. The platform was adapted to the country's linguistic diversity through a multilingual interface and designed using insights from studies on local visual symbolism, including the interpretation of icons, colors, and shapes according to cultural semiotics. The participatory design activities were led by trained local team members supported by Catalpa International, ensuring alignment with community values and practices. The involvement of education system stakeholders, from teachers and school leaders to officials within the Ministry of Education, further strengthened the contextual fit of the platform and encouraged its adoption.

In summary, adapting to the local context through a participatory design approach was key to the success and sustainability of Matenek as a management and mobile learning platform. The active involvement of different stakeholders, combined with cultural and linguistic adaptation of content, proved fundamental to ensuring the effectiveness and future continuity of these innovations in East Timor. Although educational policies remain a handicap, training school leaders, ensuring internet availability, and integrating with other initiatives were crucial aspects for guaranteeing this pilot program its long-term impact.

ACHIEVING SOCIAL IMPACT THROUGH HCI4D AND PD4D

The case studies from Nepal and East Timor analyzed in this chapter highlight the significant potential of mobile technologies to enhance education in resource-constrained contexts while also exposing numerous challenges. These findings helped us to identify insights within the opportunities and complexities of developing projects which can help secure long-term social impact.

The inclusion of local communities through approaches, such as HCI4D and PD4D, proved critical for achieving cultural adaptation of technological tools and for co-creating accurate design solutions. However, these same approaches often fall short in transferring knowledge back to the communities. While local communities may be involved in the design process, they are not always empowered with the necessary resources or capacities to autonomously continue, sustain, or replicate participatory practices in the future. This limitation underscores the need for approaches that not only include the local communities but also invest in their long-term ability to innovate and develop solutions independently.

Mobile learning technologies, particularly m-learning solutions, have clearly shown their utility and suitability as effective tools in advancing educational proposals in low-resource settings. Their adaptability, portability, and potential for offline use make them particularly suitable for contexts with limited connectivity and digital literacy. Nevertheless, the success of these interventions depends heavily on aligning them with local educational needs and cultural practices.

One significant challenge in ICT4D projects is the medium and long-term sustainability of technological solutions after initial deployment. Projects often rely on external funding, and when these funds are depleted, the big risk is that the projects could be abandoned. Furthermore, the fast pace of technological change and innovation, and also the possible dependence on proprietary solutions, intensify these issues, raising concerns about obsolescence and dependency on external providers. Open-source technologies offer a viable, affordable, and ethical alternative, allowing for greater community ownership and adaptability over time. Addressing these challenges requires projects to design and develop interventions that prioritize technological resilience and adaptability.

The durability of development projects is another critical issue. In both Nepal and East Timor, it was observed that initiatives often depend on external grants or subsidies, which, once exhausted, jeopardize the project's sustainability. This raises questions about the long-term viability of these interventions and their ability to foster community ownership of technologies. Linking the adoption of technologies with continuous teacher training programs and stable educational policies is essential. This approach would empower local stakeholders to maintain and even expand projects over time.

Additionally, the longevity of interventions is closely tied to the creation of robust partnerships. Experiences in East Timor highlight the importance of engaging government institutions, such as the Ministry of Education, to ensure continuity. Changes in national policies or shifts in funding priorities can significantly impact projects. Similarly, in Nepal, collaboration with local initiatives (e.g., OLE Nepal and offline libraries) has proven effective in embedding technologies within communities, enhancing cultural relevance and ensuring sustainable maintenance.

In HCI4D projects, bottom-up approaches are often celebrated as more inclusive and effective compared to top-down strategies. However, neither approach alone is sufficient to address the multifaceted challenges of HCI4D projects. A balanced methodology that integrates both perspectives is necessary, combining the broad strategic support of top-down initiatives with the localized insights and community engagement of bottom-up processes. A project like Kenya's M-Pesa could be closer to a middle-out approach, involving both strategic direction from providers and community-driven adoption (Markus & Nan, 2020). M-Pesa exemplifies how the interplay between strategic provider direction and community-driven adoption can successfully scale digital infrastructure in the Global South. On the provider side,

Safaricom played a central role by investing in infrastructure, strategically aligning M-Pesa with existing telecom networks, designing the service to meet local needs (especially remittances), and engaging proactively with regulators to create a flexible environment for innovation. They also built a wide network of partnerships with banks, NGOs, and retail agents to ensure accessibility and reliability. Meanwhile, community adoption was equally critical: local users rapidly embraced and adapted M-Pesa, embedding it into daily life in culturally meaningful ways, such as using it for informal savings, school fees, or emergencies. The widespread trust was supported by local agent networks who were themselves part of the community, while viral social adoption spread through kinship and remittance ties. Importantly, users repurposed the technology, contributing to its evolution and making it even more relevant. This synergy between top-down strategic initiatives and bottom-up social embedding made M-Pesa not just a technological success but a deeply rooted socio-economic infrastructure, earning it recognition as a global model for digital innovation in development contexts.

The role of Western researchers in HCI4D projects should raise critical questions about the presence of dominance of Western perspectives in these kinds of projects and their approaches. While academic research aims to achieve, among others, social justice and meaningful impact, there is a need to ensure that the needs, wants, and limitations of local communities are genuinely considered. Researchers must reflect on their positionality and the power dynamics inherent in their work, striving to foster equitable collaborations and shared ownership of the outcomes.

CONCLUSIONS

The cases of Nepal and Timor-Leste offer critical lessons on the design and implementation of mobile technologies in low-resource contexts. These experiences confirm that the transformative potential of m-learning solutions does not lie in the mere provision of devices or digital platforms, but rather in the ways they become interwoven with the social, pedagogical, and political dynamics of the communities involved. From this reflection, four key dimensions emerge to guide future initiatives in HCI4D and PD4D.

First, it is essential to recognize the non-neutrality of Western researchers and designers. Their participation in development projects is rarely impartial: it is often mediated by institutional, academic, or professional interests that may shape the direction of interventions. This reality calls for reflexive and self-critical practice, in which agendas are made explicit and an equitable distribution of benefits and responsibilities with local actors is promoted. Building horizontal collaborations based on mutual trust and transparency constitutes an ethical and political requirement to ensure the legitimacy of these projects.

Second, the identification of impact factors must stem from the intrinsic motivations of communities. Only when project objectives are aligned with local priorities—such as improving teaching, preserving languages, or managing educational resources—can communities perceive technologies as relevant and useful. Such alignment not only strengthens local ownership of the tools but also increases the likelihood of their continuity beyond the timeframes set by external funding.

Third, the durability of interventions depends on the creation of effective channels for knowledge transfer and the development of technical capacities. Long-term sustainability is not achieved merely through the delivery of hardware or software, but through the strengthening of local competencies that enable teachers, school leaders, and communities to maintain, adapt, and expand the solutions. This requires a consistent political and institutional framework that supports educational innovation, preventing projects from being subject to government turnover or the volatility of international donors.

Finally, the design of solutions must respond to the needs and profiles of educators, who serve as the key mediators between technology and learning. Teacher training in digital and pedagogical skills, along with the provision of culturally and linguistically relevant content, are essential conditions for ensuring that technological resources are effectively integrated into educational practices. Empowering teachers not only enhances the quality of teaching but also ensures that innovations remain embedded within everyday school processes.

In sum, m-learning projects in low-resource contexts require a comprehensive approach that combines cultural sensitivity, institutional sustainability, and community empowerment. Recognizing the non-neutrality of external actors, aligning initiatives with local motivations, establishing robust channels for knowledge transfer, and placing teachers at the center of the design process are indispensable steps for transforming mobile technologies into genuine drivers of educational inclusion and social justice.

BIBLIOGRAPHY

Abdelnour-Nocera, J., & Densmore, M. (2017). A review of perspectives and challenges for international development in information and communication technologies. *Annals of the International Communication Association*, *41*(3–4), 250–257. https://doi.org/10.10 80/23808985.2017.1392252

Ahmadi, A. R., Paracha, S., Sokout, H., & … (2015). Mobile mediated learning and teachers education in less resourced region. *International Journal of …. Academia. Education*. https://www.academia.edu/download/92304094/10000487.pdf

Dearden, A., & Rizvi, H. (2008). Participatory IT Design and Participatory Development: A Comparative Review. *Proceedings of the Tenth Anniversary Conference on Participatory Design 2008*, 81–91. http://dl.acm.org/citation.cfm?id=1795234.1795246

Duarte, A. M. B., Brendel, N., Degbelo, A., & Kray, C. (2018). Participatory design and participatory research: An HCI case study with young forced migrants. *ACM Transactions on Computer-Human Interaction*, *25*(1), 3:1–3:39. https://doi.org/10.1145/3145472

Holeman, I., & Barrett, M. (2017). Insights from an ICT4D Initiative in Kenya's immunization program: Designing for the emergence of sociomaterial practices. *Journal of the Association for Information Systems*, *18*(12), 900.

Kizilcec, R. F., Chen, M., Jasińska, K. K., Madaio, M., & … (2021). Mobile learning during school disruptions in sub-Saharan Africa. *AERA …*. https://doi.org/10.1177/233285 84211014860

Kizilcec, R. F., Chen, M., Jasińska, K. K., Madaio, M., & Ogan, A. (2021). Mobile learning during school disruptions in Sub-Saharan Africa. *AERA Open*, *7*, 23328584211014860. https://doi.org/10.1177/23328584211014860

López, D. C., Pera, E. M., Nocera, J. A., & Yingta, N. (2025). Designing Digital Learning Interventions for Primary School Teachers in Rural and Remote, Low- and Middle-Income Areas: A Systematic Review. In G. Bhutkar, S. Tom, D. Roy, & J. Abdelnour-Nocera (Eds.), *Designing for Tomorrow: Innovation and Equity in Global Interaction Design* (Vol. 747, pp. 125–143). Springer. https://doi.org/10.1007/978-3-032-00777-3_8

Markus, M. L., & Nan, W. V. (2020). Theorizing the connections between digital innovations and societal transformation: Learning from the case of M-Pesa in Kenya. In *Handbook of Digital Innovation* (pp. 64–82). Edward Elgar Publishing.

Matenek_-_Final_Report.pdf. (n.d.). Retrieved 8 January 2025, from https://catalpa.io/documents/3/Matenek_-_Final_Report.pdf

Ministério da Educação, Juventude e Desporto\Jornal da República. (n.d.). Retrieved 12 January 2025, from https://www.mj.gov.tl/jornal/?q=node/43

Murphy, A., Farley, H., Dyson, L. E., & Jones, H. (2017). *Mobile Learning in Higher Education in the Asia-Pacific Region.* Springer. https://doi.org/10.1007/978-981-10-4944-6

Nicholson, R., Strachan, R., Dele-Ajayi, O., & Fasae, K. (2024). Emergency Remote Education in Nigeria: Challenges and Design Opportunities. *Proceedings of the 2024 CHI Conference on Human Factors in Computing Systems.* https://doi.org/10.1145/3613904.3641921

Nunes, F., Silva, J. C., Félix, B., Melo, R., Winschiers-Theophilus, H., Bagalkot, N., Verdezoto, N., Lazem, S., Van Heerden, A., Ngubane, T., Till, S., & Densmore, M. (2024). African Co-Design: Past, Present, and Emerging. *Proceedings of the 4th African Human Computer Interaction Conference*, 316–318. https://doi.org/10.1145/3628096.3629080

Shrestha, S. (2016). *Exploring mobile learning opportunities and challenges in Nepal: The potential of open-source platforms* [PhD Thesis, University of West London]. https://repository.uwl.ac.uk/id/eprint/2962/

Shrestha, S., Moore, J., & Nocera, J. A. (2011). Open-source platform: exploring the opportunities for offline mobile learning. *Proceedings of the 13th International Conference on Human Computer Interaction with Mobile Devices and Services*, 653–658. https://doi.org/10.1145/2037373.2037476

Technology in education: A case study on Timor-Leste—UNESCO Digital Library. (n.d.). Retrieved 12 January 2025, from https://unesdoc.unesco.org/ark:/48223/pf0000387828

Thapa, D., Griffiths, D., & Kolodziejski, A. L. (2020). Constraining and Enabling Factors in the Use of ICT in Rural Schools in Nepal. *International Conference on Social Implications of Computers in Developing Countries*, 102–113.

Timor-Leste Economic Reports. (n.d.). [Text/HTML]. World Bank. Retrieved 12 January 2025, from https://www.worldbank.org/en/country/timor-leste/publication/timor-leste-economic-report

Timor-Leste—Education since independence from reconstruction to sustainable improvement. (n.d.). [Text/HTML]. World Bank. Retrieved 12 January 2025, from https://documents.worldbank.org/pt/publication/documents-reports/documentdetail/511861468761067305/Timor-Leste-Education-since-independence-from-reconstruction-to-sustainable-improvement

8 Critical AI Literacy with Young Children
In Pursuit of Ethical AI Futures

Sumita Sharma

INTRODUCTION

As artificial intelligence (AI) evolves to mimic human-like emotions, conversations, cognition, and decision-making, it is becoming an integral part of our everyday interactions. With AI employed across a variety of domains, including healthcare, education, and entertainment, children increasingly use many AI applications, often without understanding the implications of their use. Yet, AI literacy interventions with children mainly focus on explaining how things work, in child-appropriate ways, without discussing the ethical implications of the use, design, inherent limitations, and consequences of AI on children (Long & Magerko, 2020; Sharma et al., 2024, Skinner et al., 2020). Children's use and design of technology has been the focus of Child-Computer Interaction (CCI) research for decades (Druin, 2002; Papert, 1980); however, their voices, perspectives, and participation in designing ethical AI are largely missing (Iivari et al., 2024; Pera et al., 2018; Skinner et al., 2020). While ethics is a complex concept with various traditions and definitions (Durall Gazulla et al., 2024; Iivari et al., 2022; Reiss, 2010), in this work, the focus is on the ethics of design, that is, "understanding the ethical impacts of design and various digital technologies" on and with vulnerable groups, in this case, children (Durall Gazulla et al., 2024). AI systems for children are classified based on their application into four types – robots, chatbots, recommender systems, and decision-making algorithms (UNICEF, 2021), thus, CCI researchers have been delving into children's interactions with social robots (Charisi et al., 2016), conversational agents (Druga et al., 2017), recommender systems (Pera et al., 2018), and algorithmic decision-making (Durall Gazulla et al., 2024). With children's increasingly entangled and complex relationships with AI, CCI researchers have more recently focused on developing children's critical AI literacy by exploring concepts such as fairness (Skinner et al., 2020), inclusion (DiPaola et al. 2020), and diversity (Sharma et al., 2022).

CCI research shows how children are acutely aware of the disparities and inequities in their lived contexts, whether it is regarding the distribution of domestic labor, (lack of) access to education and materials, or to teachers and experts (Sharma et al., 2024).

DOI: 10.1201/9781003557777-10

Yet, they are still hopeful of technology's potential to solve many kinds of real-world problems. To thrive in the 21st century, children ought to be nurtured as future "protagonists" who understand, critically examine, and reflect on, and consequently impact the design and development of technology, including AI, in their everyday lives and be empowered to envision and drive their own technology futures (Druin, 2002; Iivari et al., 2022; Iversen et al., 2017; Iivari & Kinnula, 2018; Papert, 1980). While the Scandinavian approach to participatory design empowers and emancipates those it lends its voice to (Druin, 2002; Iversen et al., 2017), who really *get to the future* can be a matter of privilege (Howell et al., 2021). To address this gap, in the Participatory AI with Schoolchildren (PAIZ)project, children in Finland, India, and Japan explored the impacts of AI on their everyday lives and envisioned ethical AI futures in the context of schools, utilizing participatory design futuring approaches (Sharma et al., 2022, 2024). This chapter presents the approaches and outcomes of PAIZ, including children's future imaginaries, and concludes with a short discussion on future work.

CRITICAL AI LITERACY WORKSHOPS WITH YOUNG CHILDREN

In PAIZ, children explored the use of AI in their lives, discussing the ethical challenges in the design and use of such technologies, with age-appropriate examples and demonstrations. Criticality was embedded in the workshops by promoting children to question *what their future imaginaries could and could not do, and why (or why not)*, and encouraging them to envision alternative futures. This approach to criticality brought the ethics of design to the forefront, exposing underlying agendas and values, exploring alternative perspectives, and challenging dominant ideas (Iivari et al., 2022). Since these values and perspectives can be socio-culturally situated, the workshops were run in different contexts to explore diverse perspectives.

WORKSHOP STRUCTURE

Workshops comprised four phases (Figure 8.1): (1) introduction to AI in everyday life, (2) designing for the future, (3) hands-on designing and making, and (4) presentations and discussions. In Phase-1, participants were introduced to the project (its overview and expected outcomes) and to AI in their everyday life through examples of interactions with robots, generative AI (e.g., ChatGPT and Dalle2), recommendation systems (e.g., YouTube), and algorithmic decision-making. Participants also created image recognition models using Teachable Machine, further discussing the impacts of image recognition on self-driving cars. In Phase-2, participants were oriented to the future through a video explaining design futuring and were divided into small groups (with 5–8 children in each group) to envision future schools. In Phase-3, groups created a tangible version of their future designs using low-fi prototyping and arts and crafts materials, deciding what their future imaginaries can and cannot do, and why. In Phase-4, groups presented their final ideas and designs to the entire classroom, critically reflecting on their design choices, for example, *What can the future teacher do? Why? What would be fair? What would not be fair?* Each phase consisted of one to three 45-minute sessions (Figure 8.1), where there could be two

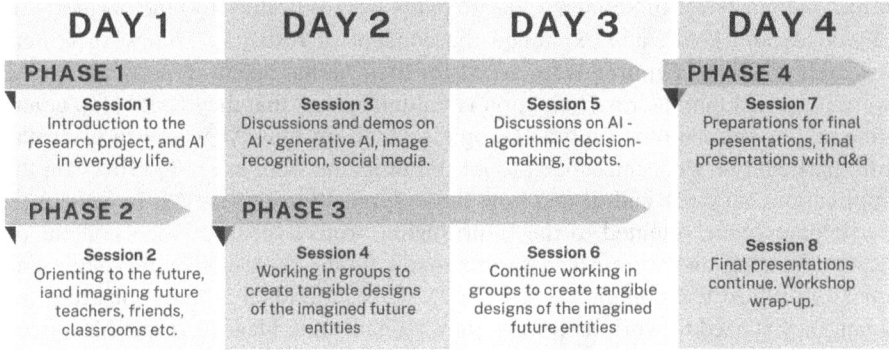

DAY 1	DAY 2	DAY 3	DAY 4
PHASE 1			**PHASE 4**
Session 1 Introduction to the research project, and AI in everyday life.	Session 3 Discussions and demos on AI - generative AI, image recognition, social media.	Session 5 Discussions on AI - algorithmic decision-making, robots.	Session 7 Preparations for final presentations, final presentations with q&a
PHASE 2	**PHASE 3**		
Session 2 Orienting to the future, iand imagining future teachers, friends, classrooms etc.	Session 4 Working in groups to create tangible designs of the imagined future entities	Session 6 Continue working in groups to create tangible designs of the imagined future entities	Session 8 Final presentations continue. Workshop wrap-up.

FIGURE 8.1 The four workshop phases, with a varying number of sessions. Each session is 45 minutes long.

or four sessions per day, depending on the time negotiated with the schools (India, Japan) or parents (Finland). At the end, participants were provided an information booklet about the workshop, activities, and links to future resources to continue their explorations of AI in their everyday lives. How criticality was embedded in the hands-on activities is described next.

EMBEDDING CRITICALITY IN THE ACTIVITIES

In each of the workshop phases, participants completed several hands-on activities with AI applications, drawing and writing in a workshop booklet (Figure 8.2), and using arts and crafts materials. Through these activities, children were encouraged to think critically about how technology shapes their experiences and the impact

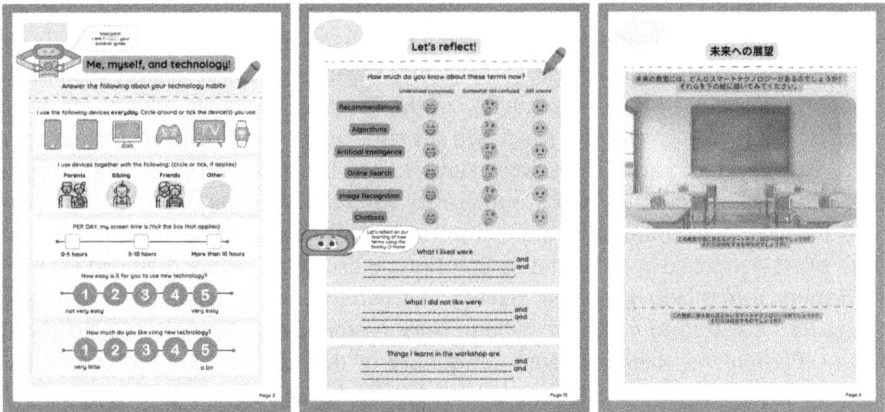

FIGURE 8.2 Pages from the workshop booklet. The English version was used in Finland and India. The entire booklet is available on the project webpage at https://interact.oulu.fi/paiz.

it has on their lives. In Session 1, participants were introduced to the project – its objectives and goals, and explained the concept of research. Following ethical research practices, children were asked for their verbal assent to participate, and were informed that their participation is voluntary, and that they could stop at any time. Using animations, such as Google's video on how AI works in everyday life, participants were introduced to AI. Participants were asked to reflect on the applications they use and, if and how those applications utilize AI. In Session 2, participants were oriented to the future using project-specific video content on design futuring and were asked to imagine a future classroom with visible and invisible future technologies (Figure 8.2). This individual activity ensured that when they started to work in groups, they all had some ideas to share. Moreover, the concept of visible and invisible technologies was discussed – that is, not all technology is tangible and physical (like desks, boards, computers), but some operate behind the scenes and might not be visible to us (like social media algorithms). In this way, critical perspectives on *what technology is?* were embedded in the workshop. At the end of this session, participants formed groups with 5–8 children in each and decided on a future entity for their future school – a future teacher, a future friend, or something else – and worked in these groups for the remainder of the workshop.

In Session 3, more specific examples of AI applications were demonstrated and discussed. Using Teachable Machine, children trained an image classifier to differentiate between two very different objects that look similar from certain angles – a blueberry muffin and the close-up of a chihuahua's face. They explored how this classifier, which only *sees* chihuahuas and muffins, *sees* the rest of the world, such as human faces and classroom objects. Participants then discussed how the classifier distinguished between chihuahuas and muffins, and what it *saw* in other objects. From this example, the discussion shifted to self-driving cars – how they perceive their surroundings, who is responsible for their functioning, and what happens if they make mistakes, such as misidentifying objects and causing accidents. Participants were again encouraged to think critically about technology and to imagine scenarios where *technology makes mistakes* and the consequences of those mistakes. Next, they explored image generation using DALL·E 2, collaboratively prompting the model and counting how many seconds it took to generate content. Discussions followed about the quality of the images, the artists who might have inspired them, who owns the generated artwork, and where such images can be used. As expected, participants were surprised to learn that AI-generated images often imitate human artists' work, sometimes without the artists' knowledge or permission. This realization motivated them to critically consider the ethical implications of AI-generated content and its appropriate use. While the Indian and Finnish workshops took place before the release of ChatGPT, in the Japanese workshops, ChatGPT was used to complete a homework assignment – writing a haiku about Kyoto. Participants then evaluated the quality of the AI-generated haiku and discussed what grade they and their teacher might assign to such work. Through these generative AI activities, participants gained awareness of the ethical issues surrounding AI-generated content and were encouraged to reflect on fairer and more responsible AI practices.

In Session 4, groups worked together to imagine their future entities, from teachers and classrooms to playgrounds and currencies, deciding what their future entities *can and cannot do*. This emphasis on what technology *cannot do* scaffolds criticality in children's future imaginaries, encouraging children to draw concrete boundaries between what is possible, ethical, fair, and what is not (Bardzell & Bardzell, 2013; Sharma et al., 2024). As groups envisioned their future entities, they first drew and wrote in the workshop booklets and then used a variety of arts and crafts materials to bring their imaginaries to life. In Finland, since the workshop was conducted in the city's digital fabrication lab, which is open to all, participants also used laser cutters and 3D printers. In Session 5, participants were introduced to recommendation systems and social media, and the impacts of those on their everyday life, with examples from research on emotional contagion and loneliness. Participants were also introduced to conversational agents, such as home assistants, and algorithmic decision-making through more examples in everyday contexts. Lastly, participants were asked to define *what a robot is* and were asked to identify what is and is not a robot from several different images, which included toy robots, home assistants, pet robots like Paro the seal, cleaning robots like Roomba, and more sophisticated robots like Boston Dynamic's Atlas and Spot and fictional ones, such as Iron Man's J.A.R.V.I.S. Discussions followed on *what is AI and what is not AI*, imploring participants to be critical of how technology is described in media or by companies, and that, not all technology is AI and not all robots are autonomous. These discussions promoted critical thinking toward technology design and use.

In Sessions 6 and 7, groups continued working on their future entities and creating tangible versions of their imaginaries. During this time, workshop instructors visited each group, asking them about their designs and what their future entities could and could not do, and why and why not. Participants were also asked to prepare a short presentation showcasing their imagined future entity to their classmates. These presentations were a part of the last session (Session 8). In these presentations, participants described and showed their designs, with workshop instructions asking questions such as *How can your future entity be fair? Be sustainable? What happens if it makes a mistake? Who is responsible?* This series of questions invited some on-the-spot debates among group members and further questions from the audience (other classmates). At the end of the workshop, participants provided feedback and reflected on their workshop experiences. Overall, through participatory design work and speculating future technologies in an educational context while being critical of those technologies, PAIZ utilized participatory, speculative, and critical design approaches for scaffolding children's visions of ethical AI futures.

PARTICIPANT DEMOGRAPHICS

PAIZ workshops were run during the school day in India and Japan, and during the autumn break in Finland, with children (ages 10–12 years). Informed consent was obtained from parents and/or the school's principal, and verbal assent was obtained from the participants after the project was introduced to them. Participation was

TABLE 8.1
Participant Demographics

City, Country	Number of Participants	Ages	Schedule	Workshop Location
Kanagawa, Japan	102 (18 groups)	11–12 years	Four days, 2 sessions per day	Three 6th-grade classes at a private school, Fall 2023
Oulu, Finland	15 (5 groups)	9–12 years	Two days, 4 sessions per day with short breaks	City FabLab (digital fabrication lab), 2–4th graders from an international school, Fall 2022
New Delhi, India	31 (7 groups)	11–13 years		One 8th-grade class at a public school, Fall 2022

voluntary, and participants could leave at any time. Participant demographics are presented in Table 8.1.

CHILDREN'S VISIONS OF ETHICAL AI FUTURES

Children's future imaginaries varied from the **probable**, including smart devices, to **possible,** including future robot teachers, to **provocative** – technologies indistinguishable from magic with unlimited storage and secret spaces. These future imaginaries embedded values inherent to children – inclusion and diversity, care and empathy, and agency and empowerment – and sought to address several different everyday challenges in children's lives.

INCLUSION AND DIVERSITY

As children imagined probable and provocative AI futures, they strongly advocated for inclusion and diversity. They embedded inclusion in these futures in terms of access to information and devices. For instance, a group of Japanese participants imagined a future textbook (Figure 8.3) that would contain all their learning materials in one place and would be so easy that "*anyone can use it*". The group was clear that their future textbook would not replace teachers, noting the significance of human teachers in a classroom. To ensure that everyone has access to this textbook, they exclaimed that it would be "*provided by the government*". Inclusion, therefore, was at the forefront of this group's imagined future, where access to the booklet and all learning materials was ensured for all children. Similarly, a group of participants in India, who imagined a future robot-teacher (Figure 8.3), explained how their imagined future teacher "*will explain our doubts very easily. This will help our (human) teachers too. She (robot-teacher) can teach us any subject*". Another group of Japanese students imagined a fashionable future teacher who uses "*high-tech equipment*" in their teachings (Figure 8.3). The group commented that not only schools and teachers, but "*anyone who wants to buy it can buy*" the high-tech learning

FIGURE 8.3   Children's future imaginaries that embed inclusion and diversity. From left to right: a future textbook (Japan), a future robot-teacher (India), a future fashionable teacher (Japan), Randoseru – magical walking stick (Japan).

equipment. Other groups imagined future robot-teachers who could teach everything to everyone in an engaging way without getting angry or scolding when mistakes are made (India and Japan). These future imaginaries make visible children's day-to-day challenges and realities when it comes to access to engaging and interactive learning materials and access to devices.

Inclusion, in terms of financial affordability, was also visible in the participants' designs, where the realities of disparities were not shunned. For instance, one group of Japanese participants imagined a magical walking stick, called *Randoseru*, that replaces schoolbags in the future (Figure 8.3). Waving this magical walking stick conjures a screen that contains all learning materials, replacing schoolbooks. When this group was asked what happens if someone cannot afford the magical walking stick, children were quick to suggest that one solution would be for that person to "*buy a cheap(er) one*". More critical forms of inclusion were also imagined and discussed. A group of Indian participants envisioned a future 3D tech-room at schools with the latest technologies to support their learning. They considered being surrounded by many types of technologies in the future, which could be combined in their 3D tech room. Technologies such as smart boards, interactive wallpaper, and an all-in-one jacket (with a water bottle, WIFI, and screens). In this grand vision, participants underscored inclusion and diversity, stating categorically how this room would be accessible to all children regardless of their gender, age, or caste, thus, being truly fair and equitable – "*3D room is fair because in future 3D room there is no gender, age, and caste inequality*". This powerful message underlines children's inherent optimism for technology to bridge inequities and showcases how their core values – of inclusion and diversity – are embedded in visions for the future.

CARE AND EMPATHY

Care and empathy are pictured strongly in children's imagined futures, from a suite of robots that help at home to robot teachers who may or may not display emotions. A group of participants in Finland imagined a suite of three robots that assisted them in the future for different types of tasks – from taking care of household work, to take care of other family members, such as a baby (Figure 8.4), "*baby bot...helps take care of babies and entertains them so it's a different type of bond*". As these robots

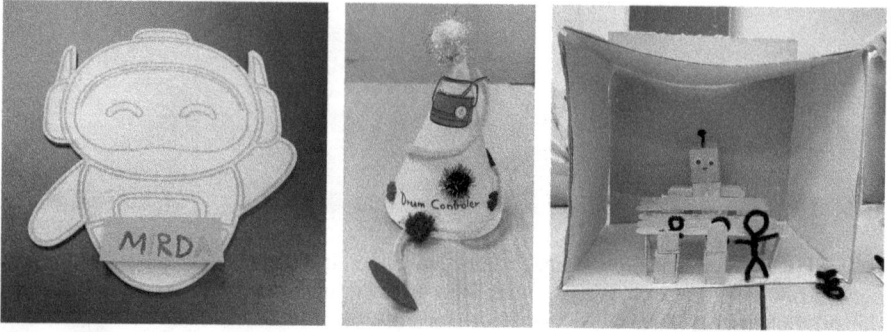

FIGURE 8.4 Children's future imaginaries that embed care and empathy. From left to right: one robot from a suite of robots that assist with everyday work and caregiving (Finland), a dream controller that ensures children only have positive dreams (India), and a robot teacher and hologram desk (Japan).

assist in caregiving and household work, they can also sense when someone is feeling down and cheer them up. Robots, and other future technologies, as empathetic helpers pictured strongly in all locations – Finland, India, and Japan, showcasing another set of values that children inherently embed in their visions of the future – care and empathy. For example, a group of participants in India imagined a dream controller (Figure 8.4) that could reduce nightmares and promote positive dreams. Participants exclaimed how their imagined future dream controller is a *"tool to have positive dreams and hence a positive effect on our well-being"*. Thus, not only physical care and well-being, through robot helpers, was emphasized in these futures, but also emotional and mental well-being by promoting positive dreams. Children displayed empathy for their future selves and their robot-helpers, mentioning that the robots *"won't do all of your chores"* (Finland), to ensure that neither humans nor robots are overburdened by the demands of daily life.

While future robots and devices were designed to spread positivity and support humans, they could not replace the need for human relationships and would be incapable of love. As a group of Japanese participants, who imagined a robot-teacher and hologram desk in their future classroom (Figure 8.4), explained that their robot-teacher *"is very good for learning, but not for human relationships"*. This group identified clear boundaries between the role of humans and robots, where guidance and advice to children at school was solely the responsibility of human-teachers. The roles of future machines were defined as helpers and supporters of humans for learning and in everyday life. Yet, there was no consensus on whether robots and AI could recognize or display emotions. Surprisingly, children imagined a variety of scenarios regarding emotions. From a robot-teacher who *"can show its emotions"* and *"is good at reading emotions"* (Japan), to robot-teachers who make *"people happy and help children in study"* but do not have or show any feelings (India). Overall, care and empathy were paramount in these futures, not only for humans but also for non-human entities supporting our well-being and work, revealing how children imagine ethical human–machine relationships, yet feelings and emotions were defined to be purely human experiences.

AGENCY AND EMPOWERMENT

Children underscored their own agency in their imagined futures – from envisioning cards and currency specifically for children provided by schools that ensure children can afford the things they need without being wasteful, to robot-teachers that provided engaging content to ensure children are not bored with learning. For instance, a group of participants in Finland imagined a future currency especially for young children that enabled them to *"buy everything"* without being wasteful, as *"the condition of this card is that you cannot buy wasteful things"*. This interesting proposal also underscores inclusion and care, in that children can afford all the things that they need, and thus, their needs are taken care of regardless of their family's socioeconomic background. A group of participants in India imagined a robot-teacher that makes learning fun and engaging, since *"nowadays people are getting bored of studying. Everyone wants new methods and ways of engagement"*. Thus, these groups imagined empowering futures where children's financial and educational needs are fulfilled through technology in the future.

Children's agency toward learning featured strongly in their designs with engaging, animated, virtual content for "everything" and with devices that teach without judgment and without getting angry. As groups categorially mentioned that their future entity *"can help us in understanding the hard concept… and children also enjoy it"* (India) and *"AI will be the teacher for difficult classes"* (Japan). Children in Japan and India also exclaimed how robot-teachers could also fill in for human-teachers *"when there are not enough (human) teachers"* (Japan) or *"whenever (teachers are) absent"* (India). Through these futures, children placed themselves in empowering positions with a strong sense of agency over their learning in terms of access to engaging and interesting materials and (robot) teachers who were always available and patient in teaching difficult concepts. Children's desired futures also uncover underlying challenges in their present realities and solutions for those challenges.

SOLVING EVERYDAY ISSUES

Children's future imaginaries addressed a variety of challenges stemming from their present realities, from heavy school bags and having access to learning materials that are engaging and available, to overpopulation and sustainable use of resources. Six groups of children (Japan) imagined future AI that could mitigate the problem of heavy school bags, from smart pencil cases and smart desks, with unlimited storage, to a magical walking stick that replaces school bags entirely. They envisioned how *"the concept of the school bag itself will disappear"*, being replaced by smart technologies, and in some cases, magical devices. One group of participants in India imagined a secret cabin that could provide space for India's growing population, explaining, *"the population is growing in India, therefore, this cabin can be used to make more living space"*, again, alluding to a magical solution to a pressing societal problem. Thus, children were mindful of their use of spaces and resources and imagined futures where such limited resources could be better utilized. Similarly, another group of participants in Japan imagined a future playground where fields for different sports are stacked on top of each other, and at any time, one is selected using a remote control. In this imagined playground, AI and robots could fill in for players if a team is short. While the group noted that such a layered playground might be expensive to build and maintain, they insisted that schools would receive help in financing it, making it accessible. In this way, schools can promote multiple sports within limited outdoor spaces and empower children to play a variety of sports.

Overall, children's future imaginaries show how they critically envisioned ethical technology futures that embedded their core values. The invisible boundaries children created between humans and machines showcase their potential to understand and critically examine complex ethical challenges in AI systems. While research on AI systems grapples with challenges for mitigating bias and discrimination, increasing trust and transparency, accountability, explainability, and responsibility, and adding guardrails to ensure humans are in the loop, many of these systems are already being used by children in various contexts. Therefore, discussions on the ethical use and design of AI systems are essential to AI literacy, especially for young children who often place a lot of trust in technology.

ZOOMING IN ON THE CULTURAL CONTEXT OF THE WORK

Through their future imaginaries and in solving the everyday issues that impact them, children reveal their socio-cultural experiences and aspirations, and the impact of those on their visions of the future. For instance, children in India and Japan sought to solve the problems with spaces – living spaces, spaces in schools for playgrounds, and spaces in classrooms. The solutions encompassed smart devices and technologies, and elements of magic – such as a magical walking stick, a secret cabin, and a smart desk with unlimited storage. Children in India imagined smart boards in their future classrooms, which they might not have access to in their classroom right now, but are familiar with, since those are already available in India in other schools. Thus, children in India imagined a variety of futures – some that are farther away (a magical secret cabin) than others (smart boards). Children in India and Japan also sought to solve systemic issues in education, with Indian children mentioning robot teachers filling in for teachers who are absent, alluding to the issue of *teacher absenteeism* in India. Japanese students mentioned the same solution for solving the problem of not having enough teachers. Thus, in their own way, children envisioned their future robot-teacher as a solution to address their everyday challenges at school.

Children in Finland also envisioned empowering futures, however, focusing on support from robots for everyday tasks, beyond school life. They did not imagine one robot (teacher) but a suite of robots helping us live our very best lives inside and outside of school. Interestingly, two groups in Finland imagined a future currency for children, which enables children to buy whatever they need without being wasteful. Thus, while all children mentioned being mindful of their use of resources, which resources they focused on (spaces, teachers, financial means) varied across cultures. Furthermore, children in Finland also explored future technologies outside of the school context, which was not as prevalent in the other locations. While this could potentially be an outcome of the differences in the workshop locations, where in India and Japan the workshops were part of the children's school day and were held in their school (classroom or lab) and in Finland during their Fall break and in the city's digital fabrication lab; it could be a reflection of young children's learned behavior of following assignments rules as closely as possible in India and Japan, out of respect for the class teacher(s) and other instructors. Thus, as expected, prominent and subtle differences between the socio-cultural context of the locations impacted children's future imaginaries. However, it is important to note that all children embedded their core values – of inclusion, diversity, care, and empathy – into these imagined futures, regardless of their cultural context.

GOING FORWARD

With children's increasingly technology-driven lives, CCI research on ethical technology design with children has started to explore alternative, diverse futures, inviting children to critically imagine their technology futures (Druga et al., 2017; Iivari et al., 2022; Sharma et al., 2024). These studies also suggest that imagining the future not only allows children to envision the world they want to create but also helps them reflect critically on current societal and technological issues (Bardzell & Bardzell, 2013). Similarly, in the PAIZ project, children's future visions tackled various everyday challenges, from robot teachers capable of personalized learning to multifunctional playgrounds that adapt to different sports, and innovative school supplies designed to eliminate the burden of heavy schoolbags. While children were inherently optimistic about technology's role in their lives, they also demonstrated the ability to critically examine its benefits and drawbacks. By involving children in the design of ethical AI futures, CCI research fosters diverse perspectives and inclusive discussions on fair and responsible technology development (Druga et al., 2017; Iivari et al., 2022; Sharma et al., 2024). Engaging with children's everyday experiences and future aspirations is crucial in fostering them as ethical designers and creators of future technologies. These critical, participatory, and speculative processes, as also employed in PAIZ, empower children to take ownership of their futures as active participants in shaping ethical AI futures. Thus, designing for the future empowers children as future agents of change, advocating for bold, provocative, alternative ways forward. Research shows that children's inherent values and future aspirations shine in these provocations, and they consider inclusion, diversity, agency, and empathy as pillars of society (Sharma et al., 2022, 2024). Designing an ethical future enables children not only to envision new possibilities but also to reflect on and address pressing challenges in their present realities, further empowering them to make the world a better place.

ACKNOWLEDGMENTS

The PAIZ project was funded by the Research Council of Finland (grant #340603) from Sep 2021 till Aug 2024. The project would not have been possible without the children and schools who volunteered to participate and energetically shared their visions for the future. The project booklets were designed in Spring 2022 by Sushmita Vavilala as a part of her final year Bachelor's project at IIIT-Delhi, India. They were translated into Japanese by Junko Tanaka, a Japanese English teacher who also ran the PAIZ workshops in Japan with three 6th-grade classes, bringing out rich and vibrant visions of the future and stories of children's everyday experiences with technology. Noura Howell (GeorgiaTech), Grace Eden (University of Tartu), and Charu Monga (IIT-Delhi) supported PAIZ workshops in Finland and India and contributed to the project in many ways.

REFERENCES

Bardzell, J., & Bardzell, S. (2013, April). What is "critical" about critical design?. In *Proceedings of the SIGCHI Conference on Human Factors in Computing Systems* (pp. 3297–3306).

Charisi, V., Davison, D., Reidsma, D., & Evers, V. (2016, August). Evaluation methods for user-centered child-robot interaction. *In 2016 25th IEEE International Symposium on Robot and Human Interactive Communication (RO-MAN)* (pp. 545–550). IEEE.

DiPaola, D., Payne, B. H., & Breazeal, C. (2020). Decoding design agendas: an ethical design activity for middle school students. In *Proceedings of the Interaction Design and Children Conference* (pp. 1–10).

Druga, S., Williams, R., Breazeal, C., & Resnick, M. (2017). "Hey Google is it ok if I eat you?" Initial explorations in child-agent interaction. In *Proceedings of the Conference on Interaction Design And Children* (pp. 595–600).

Druin, A. (2002). The role of children in the design of new technology. *Behaviour and Information Technology*, 21(1), 1–25.

Durall Gazulla, E., Hirvonen, N., Sharma, S., Hartikainen, H., Jylhä, V., Iivari, N., ... & Baizhanova, A. (2024). Youth perspectives on technology ethics: Analysis of teens' ethical reflections on AI in learning activities. *Behaviour & Information Technology*, 44(5), 888–911.

Howell, N. F., Schulte, B., Twigger Holroyd, A., Fatás Arana, R., Sharma, S., & Eden, G. (2021). Calling for a plurality of perspectives on design futuring: An un-manifesto. In *Extended Abstracts of the CHI Conference on Human Factors in Computing Systems* (pp. 1–10).

Iivari, N., & Kinnula, M. (2018, August). Empowering children through design and making: Towards protagonist role adoption. In *Proceedings of the 15th Participatory Design Conference*.

Iivari, N., Sharma, S., Gazulla, E. D., Kinnula, M., Hartikainen, H., Ventä-Olkkonen, L., & Lehto, E. (2024). HCI Design for Children. In *Designing for Usability, Inclusion and Sustainability in Human-Computer Interaction* (pp. 356–388). CRC Press.

Iivari, N., Sharma, S., Ventä-Olkkonen, L., Molin-Juustila, T., Kuutti, K., Holappa, J., & Kinnunen, E. (2022). Critical agenda driving child–computer interaction research: Taking a stock of the past and envisioning the future. *International Journal of Child-Computer Interaction*, 32, 100408.

Iversen, O. S., Smith, R. C., & Dindler, C. (2017). Child as protagonist: Expanding the role of children in participatory design. In *Proceedings of the Conference on Interaction Design and Children* (pp. 27–37).

Long, D., & Magerko, B. (2020). What is AI literacy? Competencies and design considerations. In *Proceedings of the CHI Conference on Human Factors in Computing Systems* (pp. 1–16).

Papert, S. (1980). Computers for children. In *Mindstorms: Children, Computers, and Powerful Ideas*, 3–18. Basic Book.

Pera, M. S., Fails, J. A., Gelsomini, M., & Garzotto, F. (2018). Building community: Report on kidrec workshop on children and recommender systems at recsys 2017. In *ACM SIGIR Forum* (Vol. 52, No. 1, pp. 153–161).

Reiss, M. (2010). *Ethical thinking*. In *Ethics in the Science and Technology Classroom*. Brill.

Sharma, S., Hartikainen, H., Ventä-Olkkonen, L., Eden, G., Iivari, N., Kinnunen, E., ... & Arana, R. F. (2022). In pursuit of inclusive and diverse digital futures: exploring the potential of design fiction in education of children. *Interaction Design and Architecture (s)*, (51), 219–248. https://ixdea.org/table-of-contents-n-51/

Sharma, S., Howell, N., Ventä-Olkkonen, L., Iivari, N., Eden, G., Hartikainen, H., ... & Varanasi, U. S. (2024). Promoting Criticality with Design Futuring with Young Children. In *Proceedings of the 13th Nordic Conference on Human-Computer Interaction* (pp. 1–15).

Skinner, Z., Brown, S., & Walsh, G. (2020). Children of color's perceptions of fairness in AI: An exploration of equitable and inclusive co-design. In *Extended Abstracts of the CHI Conference on Human Factors in Computing Systems* (pp. 1–8).

UNICEF (2021). Policy guidance on AI for children version 2.0. Recommendations for building AI policies and systems that uphold child rights. Retrieved on January 15th, 2025, from https://www.unicef.org/innocenti/reports/policy-guidance-ai-children

9 Designing an mHealth System for Non-communicable Diseases in Egypt[*]

*Galal H. Galal-Edeen, Amit Gudadhe,
Po-Ying Chao, Sarah Suib, Luc Bolier, and
Ahmed Sorour*

BACKGROUND

Egypt's out-of-pocket (OOP) spending rates on healthcare is extremely high by international standards, estimated to exceed 60% of total healthcare spending (Fasseeh et al., 2022).[1] A brief from the Egyptian Curative Organisation[2] (one of the organisations that comprise the Egyptian state healthcare system), emphasised the need for a non-communicable disease (NCD)-focused system offering patient education and national data collection. Mobile-based solutions were seen as a promising direction.

A meeting with the main client was held, and identified a number of ideas for a health-care support system using smart mobile phones with the following potential objectives:

1. Distance case management: The idea of using a mobile phone to access full patient records and exchange radiographs for diagnostic purposes has been ruled out.
2. The discussion identified a potential for using mobile applications in the area of NCDs (such as hypertension, diabetes and high blood cholesterol). Two broad categories of needs & functions can be identified:
   a. Patients normally have poor access to basic information and advice to manage and stabilise their conditions; there is a need to provide information on preventative measures and healthcare advice to patients.

[*]This project has been funded with support from the European Commission. This communication reflects the views only of the author, and the Commission cannot be held responsible for any use which may be made of the information contained therein.

DOI: 10.1201/9781003557777-11

    b. There is no national database on disease incidence and characterising features. There is a need to have a central record of patient demographics and disease incidence and characterising features to guide national healthcare policies.
3. There is a potential for applying for World Health Organization (WHO) funding.
4. This project can be integrated with the National Strategy for NCDs 2010/20. The document "National Egyptian Initiative for The Prevention and Control of Non-Communicable Diseases" was cited.

The team was based at the Faculty of Computers and Information at Cairo University.[3]

CULTURAL AND RESEARCH CONTEXT

Egypt is one of the most populous countries in Africa and the Middle East. The country has a population of over 107 million (CAPMAS, 2025a), with the capital, Cairo itself, being home to around 20 million residents. The literacy rate is around 74.5% for adults aged 15 years and over (CAPMASS, 2025b). Arabic is the main and official language, although basic English is spoken in the capital city (Cairo) and in most major cities.

Egypt's strong family-oriented culture and high population density present both opportunities and challenges for healthcare delivery. High rates of smoking and sugar consumption (in both food and drink), along with limited access to reliable health information, compound the burden of NCDs. Illiteracy and traditional norms required field researchers to adapt their approach, using visuals and culturally appropriate materials.

The people are, in general, courteous, helpful and accepting of others. Egypt has seen a multitude of incoming immigration waves throughout its long history, from Africa, the Middle East, Asia and Europe. The country has a strong cultural tradition based on the family as the central unit. A high level of family cohesion is evident, with members of the family committed to helping each other, which explains why many extended families tend to live within a close geographical proximity of each other. In general, the Egyptian population has a high rate of tobacco consumption, such as cigarettes and shisha (a water pipe for smoking). Around 35% of the total male population are smokers (CAPMAS). Moreover, smoking indoors is allowed in many places, although this is being visibly reduced through extensive smoking bans in some institutions; designated smoking areas in cafés and restaurants are currently seen to be on the increase.

FIELD RESEARCH

The team conducted interviews with patients, doctors and officials to answer key questions about lifestyle, healthcare access and technology use. The target population—low-income groups with limited education—often lacked awareness of their conditions and relied considerably on family for support.

THE CULTURAL SIDE

There are social norms and traditions that field researchers must be sensitive to, like the implicit rules of conduct in public, showing affection, attire and approaching members of the other gender without prior introduction. Since the team had members who were culturally strangers to the research context, approaches to field research needed to be adjusted accordingly, so special project and personal introduction leaflets and cards containing text and team photos were produced to facilitate access. See Figure 9.1.

The first phase of research was explorative in nature, aiming to map out the peculiarities of the research space. Egypt is a high Power Distance (PDI) country (Hofstede, 2011), so several permissions had to be obtained, after which various initial visits to the hospitals, and interviews with several physicians, public and private, as well as a meeting with officials from the Ministry of Health and Population (MoHP), were conducted and the team was able to finalise the following set of research questions:

1. *What kind of lifestyle do people (at the BoP) have?*
2. *What is the impact of an NCD on a patient's life?*
3. *How does the current healthcare system of Egypt work?*
4. *What technologies are available to the BoP?*
5. *What is the professional opinion of the healthcare system and NCD in particular?*
6. *What are the needs, values and future hopes for the BoP?*

FIGURE 9.1 Team introduction leaflets.

These research questions required a variety of audiences to answer them, as well as a variety of research tools. The team used a mix of literature reviews, field observational techniques, interviews and meetings in which respondents collaborated in building a picture of the topics being investigated.

The Target Group

The group targeted by this intervention is a Base of the Pyramid (BoP) group, typically characterised by being at the lower spectrum of the national household income levels, with a level of education that is a little more above the literacy level, which impacts their knowledge of their diseases and access to information sources in general.

Egypt suffers a high rate of NCDs such as hypertension, diabetes and high blood cholesterol. NCDs are chronic, non-infectious diseases that generally exhibit slow progression but have a high impact on the patients' quality of life. Figure 9.2 shows a distribution of NCDs as of 2011.

Cardiovascular diseases are the leading cause of death in Egypt, with NCDs accounting for a significant portion of the disease burden, with significant geographical disparities in healthcare access (Radwan & Adawy, 2019).

The Technology Side

The rate of mobile communication diffusion is much greater than that of computing technology in domestic settings. At the time of the start of this project (2011), there were 74.44 million mobile subscribers in Egypt; the number now exceeds 116 million, with smart mobile phone users estimated at 71.24 million (source: https://www.statista.com/statistics/748053/worldwide-top-countries-smartphone-users/). This level of penetration presents a great opportunity to increase services via mHealth applications (healthcare applications on mobile) for those in need. That was

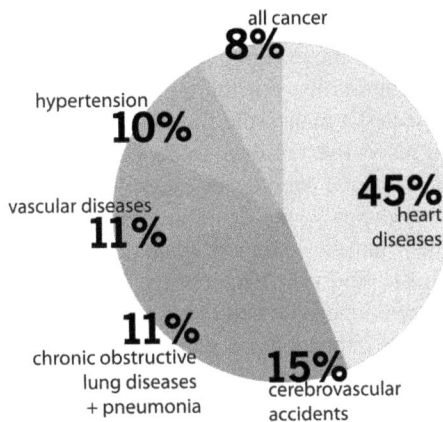

FIGURE 9.2   Distribution of NCDs in Egypt (2011).

precisely the opportunity spotted by the high-ranking MoHP official from whom the team took the triggering idea.

IN THE FIELD

The contemplated intervention is somewhat culturally invasive, as it requires the involvement of a wide variety of people in the health aspects of the BoP of Egyptian society, with its not-always-obvious, peculiarities. This situation required a foundational understanding of the culture and the prevailing norms, attitudes and values.

After spending some time in the field, a better image of the interviewees emerged, and the research tools were adapted. For instance, the team brought workbooks to the hospital, but people could not read them because of the high level of illiteracy encountered. The team decided to use the workbook as an interview tool instead, so the questions were read from the workbook, and the responses were written down or drawn. Also, the pilot tests showed that the workbook did not garner as much information as hoped for, even from literate people. Other interesting insights emerged; for example, that people liked to talk about family and friends, they were not all able to read, but they loved photos and images. With this knowledge in mind, and after the team attended a seminar by Dr Apala Lahiri, from Human Factors International, on conducting research in developing markets, the team was inspired to develop more suitable tools for patients and BoP people. The amount of text used in the research tools was reduced to a minimum, and images were used to symbolise things. This made it much easier to communicate with interviewees from the target population. The team also reduced the amount of written text on the research tools, like the workbook and the flash cards.

KEY OBSERVATIONS

LIFESTYLE AND NCD IMPACT

Participants were asked to map their daily life. It was noted that most people had quite a consistent daily routine. Women in general take care of the family and the housekeeping. In the afternoon, they spend time with their children or with friends around the neighbourhood. In the evening, they prepare meals and have dinner, and watch TV together with the family. For the men, more often, work was on a daily basis. The team also detected a major role for their social networks. Men also liked to spend time with their respective families. In the evening, they have dinner together and watch TV. The importance of people's own family was further established during interview sessions with patients in the hospital. Families are often large and traditional. People rely on their families' support when they get sick. The house of a family is a very personal space, and visiting people at home is not too common. While interviewing doctors, they pointed out that house visits were therefore very unusual. Most people at the BoP are dependent on public transportation, which is relatively cheap within Cairo (although the cost of transportation has been rising steeply more recently). However, the team observed and experienced that moving through the city of Cairo can be time and energy consuming. This puts personal and economic burdens on patients who choose to go for personal health-related consultations.

Most patients reported that they discovered their disease by accident, sometimes at a late stage. Once being diagnosed with an NCD, most patients did not mention many changes in their lives. If they did not feel sick, they often did not take any measures against their disease.

Patient remarks:

> **I don't go to clinic often. I go to big hospital because they have good doctors.
> I only go to FHU to get some information, sometimes.**

Patients suffering from NCDs were asked to map their world to find out (Figure 9.3), among other things, what medical facilities they visited. Most patients visited mainly the large hospital in the city and did not often go to primary health units, which were closer to their homes. One of their reasons for this is that they think that the large hospital has better-trained doctors and facilities. Also, the large hospital provides free medications, whereas local pharmacies mostly charge money.

THE HEALTHCARE SYSTEM AND ITS USAGE

As in most countries, Egypt has a few large hospitals providing secondary care. Spread throughout the country, many primary health care facilities are available, called Family Health Units (FHUs). The FHU has many basic facilities to provide treatments for general diseases, early detection and monitoring. Health care as well as basic medication is provided free of charge at any public health facility. In the city

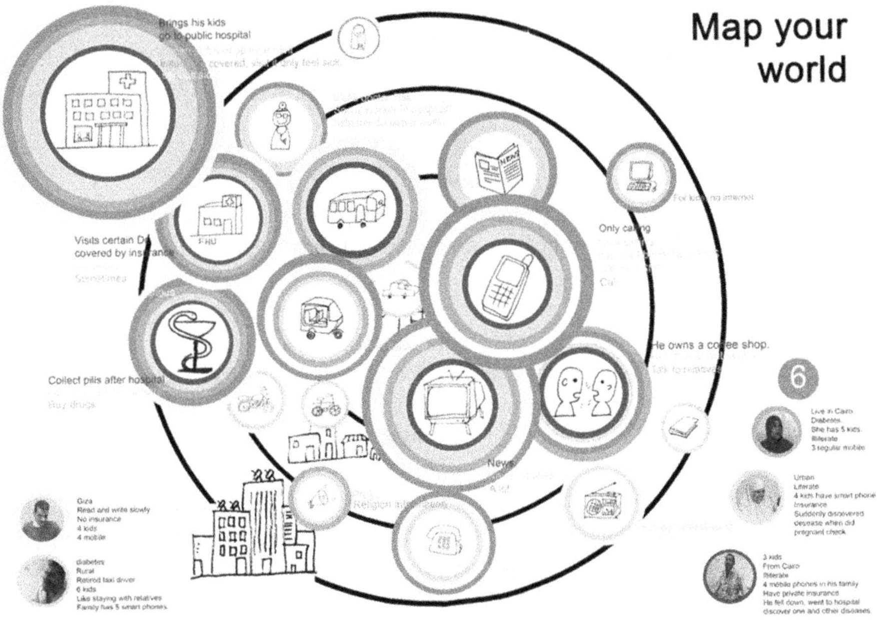

FIGURE 9.3 Patients' mappings of their world.

of Cairo, a huge number of pharmacies are operational (under State license). Those pharmacies provide certain basic services such as blood pressure measurement and blood sugar level tests; a registered Pharmacist must always, by law, be present. In general, patients do not have a fixed General Practitioner (GP) as is the case in some European countries, such as the United Kingdom and the Netherlands. Each time they visit a health unit, a different doctor might see them. Raeda Refiya (RR) (which roughly translates to "Rural Adviser", but we'll use the term "Health Visitor" here) are social health workers spread across the country, primarily covering rural areas. Their job is to carry out home visits within their community and spread health information, particularly on family planning. They mainly talk to women in the age range of 15–49, each of them covering around 500 families. They receive and report health information to the local FHU. They refer people to the particular FHU when they think a person should see a doctor.

The MoHP organises periodic training for the RR to provide them with the right level of knowledge. They sometimes distribute health information to the families they're assigned, which can be visually more easily understood due to the high rate of illiteracy prevalent in BoP communities. However, in patient interviews, it was noticed that most were not aware of the existence of the RR.

HEALTHCARE SYSTEM LIMITATIONS

Because public hospitals cannot refuse anyone, it's often the case that the number of patients exceeds available capacity, resulting in less care for each patient and sometimes in the need to share a bed among two patients. This is one of the reasons why most people who can afford it prefer going to a private healthcare provider rather than a public one, despite the significantly higher cost. From a questionnaire completed by students, it appears that most of them only make use of private health providers. There is general distrust in the service of the public health provider. Some patients made remarks like:

> *I don't trust any government system*
> *You might die in a government's hospital even if you're not sick*
> *I feel that doctor is very busy and does not have time to look after me, but he is good.*

Doctors working in a public hospital all stated that the quality of health care was very high. They did mention that they always lacked sufficient capacity and that nursing staff were often overwhelmed due to the high number of patients they had to deal with. When asked what should be improved in the current health system, both doctors and patients stated that there is a need for better funding. From the doctors' perspective, they currently cannot generate any income based on treating a patient. Also, providing a patient with health insurance could reduce the burden on the public hospitals, according to them. People mostly do not have health insurance, nor do they consider paying for one. Sometimes their employers provide health insurance schemes for them. They think that insurance could improve the service they get and would allow them to go to other (private) providers, rather than the large public hospitals, where patients think the quality of the healthcare should be improved. Doctors stated that the lack of health knowledge and awareness among patients was a major factor that needs to be improved. They also state that a doctor's time efficiency could be improved if patients' health records were easily available and communicated between doctors and departments. Figure 9.4 provides a comparison between how patients feel about the public and private healthcare systems.

TECHNOLOGY ACCESS AT THE BoP

In order to investigate what main technologies people at the BoP have access to, we asked patients in the hospital to point out which communication tools they used mostly. Patients were asked about whether and how they used their mobile phones. Most of them owned basic mobile handsets, mostly used for phone calls. In some cases, each family member had a personal phone. Although most of them were illiterate, they did sometimes use text messages with the help of their children.

One patient remarked:

Sometimes I need to ask someone to help me to dial numbers.

The patient also reported that they watched a lot of television. To see the trend of mobile technologies in Cairo, the team asked students about their mobile phone usage. It appeared that smartphones are becoming popular, and several of the participants used 3G/4G mobile internet, which is widely available in the city, with mobile telephony operators currently (in 2025) announcing the rollout of 5G networks. When asked what communication channels would be best to receive health advice and information, the patients often choose the television. They also pointed out phone calls with professionals and advice from family and friends. Doctors had various opinions, but mentioned that family and friends cannot always provide the correct advice. Rarely do patients choose home visits as a solution for obtaining health care, and usually, when it is very hard for the patient to be transported to the clinic. However, going to a private clinic often involved significant cost, transportation and waiting time burdens. Students, besides TV and family, often choose advice from professionals via telephone or health events as the preferred channel. Figure 9.5 illustrates this.

NEEDS, VALUES AND HOPES

Based on contact with patients, doctors, health workers and students, the team identified their needs, values and hopes for the future in a broad sense. Doctors pointed out that they would like to see better health education for the public. They all agreed that patients lacked personal responsibility and awareness of their own health situations. If they were informed better at an early stage, the number of people getting sick could be greatly reduced. One reason they give for their patients' behaviour is the often-poor conditions of their daily life, which do not allow them to sufficiently care about their wellness. Their own work could be made more efficient if they could keep better track of patients' health records and share them

Private

reality

bunch of theifs trying to get as much as they can

better than public healthcare

convenient

available most of the time

shorter waiting list   affective and speedy

faster   quickly

faster

feeling

trust it   care about patient   "deal with the patient as human"

doctor's appear as more experience   better treatment.

cases are handled in a better way   clean, trusted, accurate

professional   Good service   suitable money

good CARE to patients   big caring   More care

More guaranteed as regards safety and protection

environment

best medical service   very clean   better quality

more safety

new and well maintained and well calibrated medical devices.

technological techniques and machines in handling the cases

Public

reality   the best doctors

population increases public healthcare becomes less effective.

needs more time and more effort to be efficient and effective

does not care

feeling

public is not good

i dont trust any government system

"you maybe die in government one even u if u were not sick"

convenient   cheap

the system supported in my organization

good public hospitals open only till 2 p.m

environment

medications and equipments to the staff and nurses are in misreable shape.

government support   lack organization

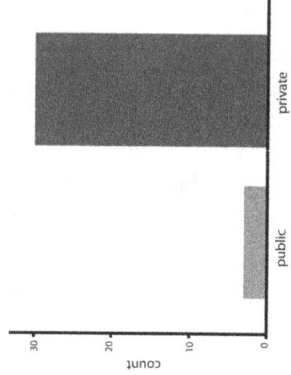

FIGURE 9.4 How do people feel about public and private hospitals.

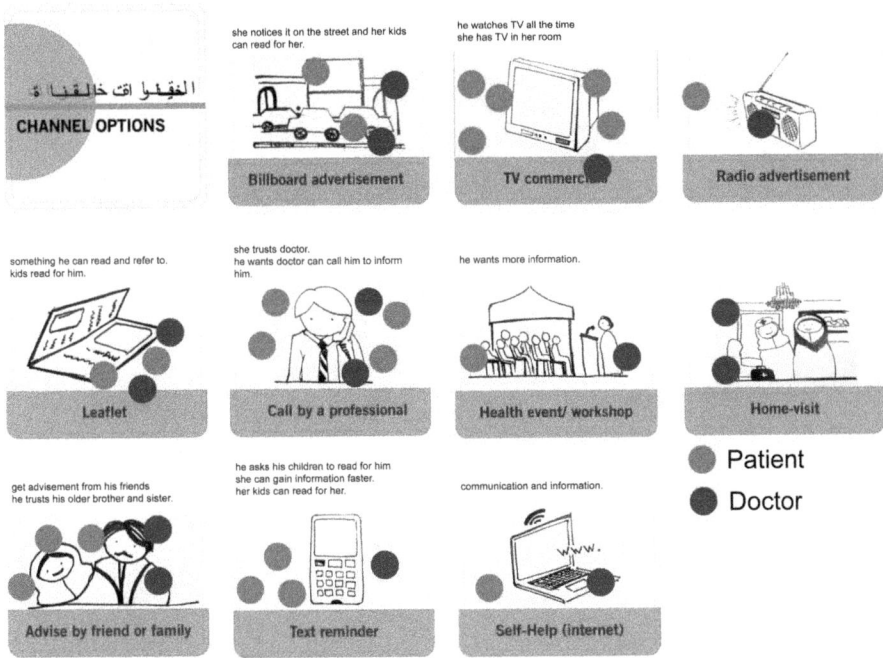

FIGURE 9.5 Healthcare access channel options.

easily with other doctors. Also, a way to encourage the patient to show up for follow-up meetings could be helpful, since most patients do not return after treatment until they become (very) sick again.

The team distinguished three types of information coming from healthcare providers to patients. The first is educational information to stimulate a healthier lifestyle. The second one is awareness in terms of early screening and detection of symptoms. The third is the medical advice given directly by doctors. Figure 9.6 is a schematic of the information flows from and trust level in the current public healthcare system in Egypt.

STAKEHOLDER INSIGHTS

Most of the people using public facilities belong to the BoP. From the research, the team concluded that the health information and awareness of the BoP are not sufficiently addressed. Having better information available to them and making them more aware of their health issues could reduce the development of NCDs. Patients that already suffer from an NCD could benefit by taking more action to suppress the consequences. The team decided to focus the application on improving awareness, early detection and monitoring of NCDs. Based on this focus, the team created summaries of the perspectives of all stakeholders involved.

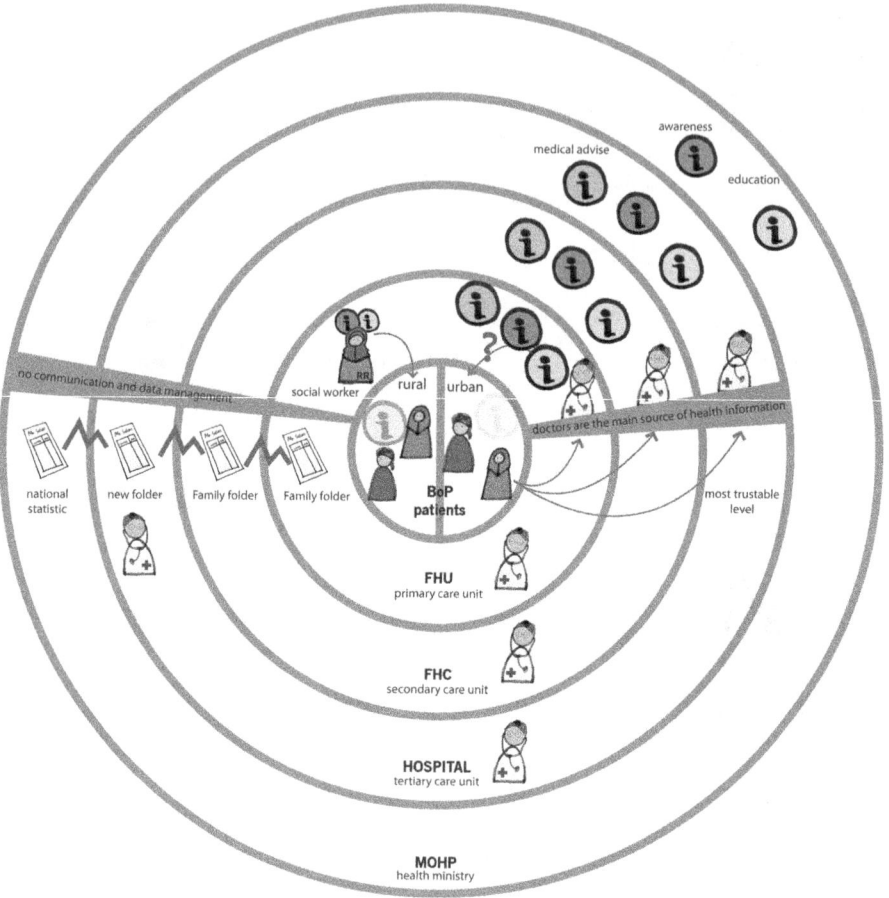

FIGURE 9.6   Healthcare system information flows and the high trust level.

Ministry of Health and Population (MoHP)

There is a need for the MoHP to gather proper disease demographics, which will aid in formulating a better healthcare strategy for Egypt's healthcare system. Insights into the health of the population allow for better allocation of resources.

> *Currently there are no national health registry systems that can provide us with proper NCD statistics.*

Dr Refaat, MoHP

Doctors

Doctors believe that a lack of information, awareness and responsibility among patients increases the amount and severity of diseases. If patients show up earlier and more often, they could be treated better and more efficiently. The lack of resources in the hospital adds to the inefficiency in treating the patients.

It is a challenge for me to diagnose patients since there is no documentation on patient's current diagnosis and medication. It is the patient's responsibility to keep the documents they receive. If I had access to a patient's previous health information, it could save me a lot of time and I would be able to help my patient more efficiently.

Dr Ahmed, 27, Kasr Al Ainy Hospital (a major State hospital, the biggest and best-known hospital and medical school)

Patients

Most people do not know how to obtain health information or do not have a trusted source apart from their visit to the doctor. It was also observed that the patient follow-up rate in Egypt is very low. Many people visit the doctor only when their situation is really bad; culture, finance, work and illiteracy are some factors contributing to this. Some interviewees explained that they stop their medication as soon as they start feeling better. The people do not currently understand the importance of aftercare and the monitoring of their diseases.

I do not know where or how I can get more information about my disease.

Fatima, 40, Housewife (a Diabetic)

Health Visitors (Raeda Refiya)

The RR are women assigned to educate the public in the rural areas about a healthy lifestyle. They generally operate in their own community where they are well-known. With a minimal education level (high school), they are trained and employed by the MoHP directly. Presently, they visit households, targeting the women and imparting information on topics such as pregnancy, childcare and family healthcare. However motivated, the role they currently play to increase people's health is limited, as well as their reach. The RR does not operate in urban areas.

I want to strengthen my role in the public healthcare system and raise my status in the community.

Sanaa, 43, Raeda Reyfiya, Giza

Pharmacists

A large number of pharmacies are spread across Egypt, even in rural areas. They are easily accessible and owned by mostly private but also public sector operators. Every pharmacy in Egypt must be attended by a certified Pharmacist by law. The pharmacies currently provide free Blood Pressure and Blood Sugar Level tests for a low or no fee. It was noticed that some large chains also keep in touch with their customers and provide extra benefits to regular customers.

Pharmacy business is a highly competitive market. I am always searching for new means to increase my sales.

Mahmoud, Owner, 35, Al Noor Pharmacy

DESIGN FOCUS AND RATIONALE

Based on the information assimilated during the research phase above, the team generated initial ideas that would best suit the context revealed through the fieldwork. Firstly, the team generated criteria that had to be met by the design concepts. These criteria were developed based on involved stakeholders, findings from the research, the available skill sets of the project team of designers and engineers as well as what the team learned from similar projects and the relevant literature. Divided into five categories, the following requirements and criteria were determined:

1. NCD statistics
   - The MoHP should be able to acquire data on how many and where people suffer from NCDs (disease demographics).
2. Innovative aspect of the directions
   - Improve the efficiency of the current paper-based health management system.
   - Supply an informative system to the BoP people and the NCD patients.

3. Sustainable system
   - Ability to be scaled up.
   - Added value to the current service.
   - Financially sustainable (a system based on trade, rather than aid).

4. Benefits received and the additional value it provides to people at the BoP
   - Increase their communication touchpoints with the healthcare providers.
   - Increase awareness and responsibility for their health.

5. Application-based system
   - The system to be developed should make use of the extensive availability of mobile phones.
   - The system should be a programmable application for (mobile) electronic devices.

With these aspects under consideration, the team ideated in different directions. Since the form of the concept is an application, the team decided to pick one stakeholder as the main user of the application. The team came to the conclusion that there were two potential directions: taking either the RR, or the Doctors as the main user of the application. The following pages present the potential directions considered.

WHY ARE PEOPLE AT BOP NOT THE CENTRAL USERS?

1. *The issue of illiteracy*
2. *Cultural and lifestyle factors*:
   - *Low personal responsibility for one's health.*
   - *Priority in life is more about survival than health.*
   - *Challenges to integrate any new idea in the current system/situation.*
   - *Need a trustworthy source to convince them of a new system.*
3. *Unreliable source for NCD statistics*
4. *Limited sources of (mobile) technology*

DIRECTION 1: Focus on Raeda Refiya

Description

This idea aims to empower the RRs by increasing their role in the system, by providing them with better means of management of information and patient data. It also aims to increase the trust among the community in the need for and importance of RRs.

The basic thought behind this direction is to tap into the existing structure of the RR system in the rural areas and use existing available mobile phones. During the first visit, the RR will register the family members via her mobile phone (step 1). Patient's data and previous medical records will be entered (or possibly transferred). Based on this entered data, the system reorganises individuals into various categories like priority and normal. Also, based on the patient's state, the frequency of visits will be decided by the system (step 2). This will help the system to generate weekly visit schedules for the RR, i.e., families they should visit in their assigned area. The system can transmit basic health info on pregnancy, NCDs, general health, etc., which may be retrieved by the RR (step 3). Also, the MoHP can pass on important information to people, like outbreak of diseases or the organisation of NCD camps, via this system through sending update messages to RR. This would be efficient and helpful when visiting the families, as compared to the present scenario, where they generally do not take any documents during a visit. Every time a file is updated, the NCD database also gets updated. In this scenario, the doctors at the nearest FHUs are also included in the loop, as they are the means to obtaining reliable NCD diagnoses.

DIRECTION 2: Focus on Doctor

Description

The first time a patient visits any public healthcare provider (hospital/Family Health Centers/FHUs), he/she is registered at the reception based on his/her National ID number. Every time he visits the doctor, the doctor updates the patient's file through his smartphone application. Various prelisted parameters like diagnosis, blood pressure and blood sugar levels, and follow-up dates make this a fast process. Once the patient is discharged from the health centre, he keeps receiving regular health information, follow-up reminders and lab test reminders based on the health programme he was registered for by the doctor. This information is sent by the central server and is available in the form of both text and pre-recorded voice calls. This idea also utilises the existence of numerous pharmacies across the country. Each pharmacy provides free blood pressure and cheap blood sugar tests carried out by an authorised pharmacist. This facilitates the system to penetrate deeper sections of the society by providing a cheap and easy access point towards monitoring patient data. The main advantage for the doctors is access to the patient's medical history, which is currently unavailable, leading to better diagnosis and treatment. Another advantage is that doctors are highly educated, have access to and are easy adapters to new technologies.

Evaluation

Based on the criteria, we evaluated the two directions in order to decide which one to develop further. Both proposed directions are very interesting and pose a number of advantages as well as disadvantages. In fact, from a futuristic point of view, both directions could be combined to form one robust system with both sub-systems complementing each other. However, implementing a combined system right away would pose many complications in the present scenario. A first change in the current healthcare system should be on a smaller scale, evaluated and may be further expanded as time passes.

Based on the criteria, potential directions, comparison of the concepts, as well as project requirements and discussions with the client, the team decided that the second direction (the Doctor as the main user) would be best suited to be pursued further in the following phase. The reasons for favouring this choice may be briefly formulated as:

- *The doctors are highly educated professionals and also have a greater affinity with new technologies like smartphones.*
- *Higher availability of reliable statistical data.*
- *Better disease diagnosis by the doctor due to the ready availability of the previous patient history.*
- *Improving the image of the healthcare system by offering better service to the patients.*
- *Possibility of creating a financially sustainable system.*
- *Potential for pilot introduction and afterwards to scale up the project.*

Figure 9.7 shows the evaluation matrix, as completed for the two alternative foci directions.

The Aafia System Prototype
Named after the Arabic word for "good health", the Aafia system supports interaction between doctors, patients, health visitors and the MoHP. It offers:

- Digital patient registration and history tracking
- Automated reminders for follow-ups and lab tests
- Educational content tailored to individual conditions
- Data sharing among doctors and health units
- Centralised NCD data collection

The system is designed to facilitate interaction between four different stakeholders through a multifaceted platform service system. The system will serve different users simultaneously in order to create different values to fit their respective needs. Figure 9.8 illustrates this.

Each user will experience a different service through the system. These service touchpoints highlight the scenario of how each stakeholder engages with the service system and how it will create an impact to improve the overall "Aftercare" service and help gather required statistics. This is illustrated in Figure 9.9 a–d.

The system comprises five layers of functions at its core in order to create a holistic environment to fit the needs of the different users identified. Figure 9.10 displays the different functions: the coloured sectors represent layers beneficial for each stakeholder. The arrows denoting the data flows between users and the "Aafia" system show what data the users need to provide to the system and what they can gain as a result.

Evaluation

Concerned stakeholder	Direction 1: RR as main user			Direction 2: Doctor as main user			
	RR	MOHP	BOP	Doctor	MOHP	BOP	Pharmacy
Technology used	Not familiar with new technology and need smart phone	Set up server and computer used	SMS audio call	Easily adaptable to Smartphone app and most of them have smart phone	Set up server and computer used	SMS audio call	Easily adaptable to Smartphone app
NCD statistics	Cannot provide accurate number as they are unable to diagnose	Can make extra medical convoys to certain square		Provide more accurate number from diagnose	Data only on those who approach the public health service		
Health info	Info can reach family by established network for RR	Can spread info to target group	Reach only women of the society	Detailed info on prev diagnoses and advices	Can spread info to target group	Better reach to healthcare information	
Healthcare	More involved in healthcare system (greater role and status)	Build new hospitals according to needs of the people	No improvement on healthcare other than prevention	Keeping good track of discharged patients	Build new hospitals according to needs of the people	Better aftercare	More connection between BOP
	Can access people easily	Built connection between FHU and family via RR	Limited knowledge of the RR. less confidence to the people	Potentially more time spent on documentation	More efficient healthcare system	Only valuable when visiting a hospital (in comparison with the other concept)	
	RR unable to comprehend the seriousness of some cases	Improve the status of health units in villages	Information only covers the rural areas	Privacy issue			
	Not highly educated may not have a bachelor degree	Decrease number of people coming to big hospitals like Kars Al Any		No special function will be done by the app except documenting			
Business model	Need funding with indirect return				Potential sustainable business model		Potential sustainable business model

Blue: Advantage   Red: Disadvantage

FIGURE 9.7 Design concept development.

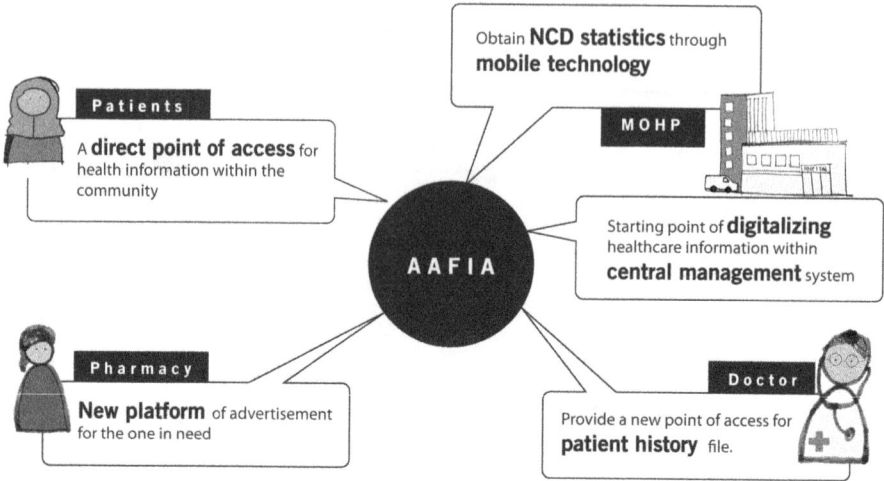

FIGURE 9.8   Services and involvement of stakeholders.

Service Ecosystem and Value Offered
The service ecosystem highlights the process experienced by each stakeholder and how the system facilitates their mutual interaction. In Figure 9.11, coloured arrows indicate the data exchange and process steps dependencies between different users and the Aafia system.

PROTOTYPE DEVELOPMENT

An Hyper Text Markup Language (HTML)-based prototype was developed and tested using Axure and Android. Feedback led to changes in interface flow and the inclusion of key medical data fields. User testing confirmed the value of simple visuals and voice-based notifications for BoP users.

Communication between Designers and Programmers

The team created visuals on paper of the different screens of the application. The programming team developed use case models, including all functions for each stakeholder separately. This model could be used as input for the programmers and could still be understood by the designers.

An early HTML-based working prototype was developed using the Axure RP tool which was easy to use for both designers and programmers (Figure 9.12). This was also used to do some initial user tests without the need for a smartphone, since the application should also run in any of the popular internet browsers.

The Application
This application is to be used specifically by doctors during their work in the healthcare facility. The doctor is the most important stakeholder for this part, and the design of the application shows the possible interaction between the doctor and his/her own phone. The application prototype was tested on Android-based smartphones.

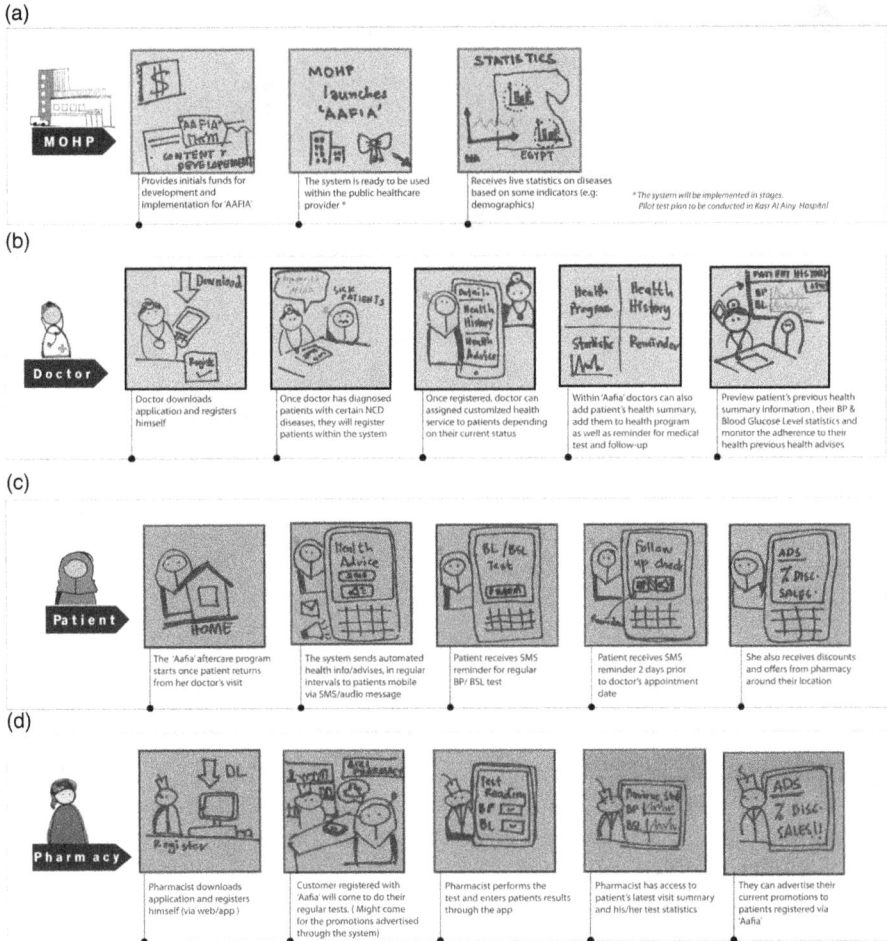

FIGURE 9.9   (a–d) Scenarios of stakeholders' engagement with the designed system.

The application provides time savings for the doctors in the hospital. As the team discovered during the field research, the need for better documentation of patients' health records is high. This is especially useful when the information can be shared between different doctors in different departments or hospitals. In the field research, we also found that doctors were having difficulties with the low level of knowledge that patients had about their own health and their diseases in general. The application gives a possibility for the doctor to provide the patient with specific health information that concerns him or her. Also, reminding the patient of scheduled follow-up meetings can reduce the large number of patients who currently do not show up for follow-up appointments. The application is focused on patients suffering from diabetes and hypertension, which are very commonly found among the Egyptian population. This focus makes it possible to easily fill in and review the important data for those diagnoses. After a successful implementation, the application can be customised for doctors in other fields.

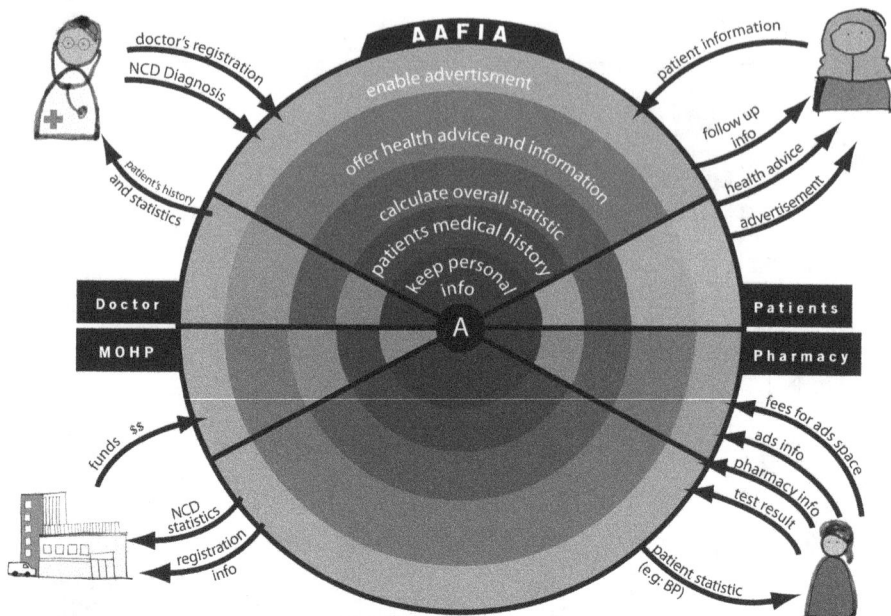

FIGURE 9.10   Data Exchanges Between Stakeholders and the Designed System.

Initial User Tests

We used Axure software to develop a fast prototype in HTML-code. Some user tests were performed over the internet, observing the participants' behaviour by asking them to share their screens using a Skype connection. The participants were asked to read and follow instructions, giving them tasks to accomplish.

The team received valuable feedback both on users' understanding of the interface, as well as medical knowledge concerning the function of the application. The order of screens (interaction sequences) was modified radically after the test. Also, some buttons were changed in order to improve the understanding of their function. Suggestions from a medical perspective to improve the application were taken on board. A "must" data item was information about the body weight of the patient. Another one was to add the medication history of the patient. Patients often do not remember the names of the medications they are taking. The graphical displays were found to be very useful.

FUTURE EXPANSION

Potential expansion in the future within the core functions established in the first phase, the Aafia system can later be extended to benefit more potential users:

- RR can be integrated within the system (similar to Direction 1, as discussed earlier), creating a new service system within the already built-in database. RRs can benefit from the personal information on patients as well as access to healthcare. And the medical information database is currently unavailable

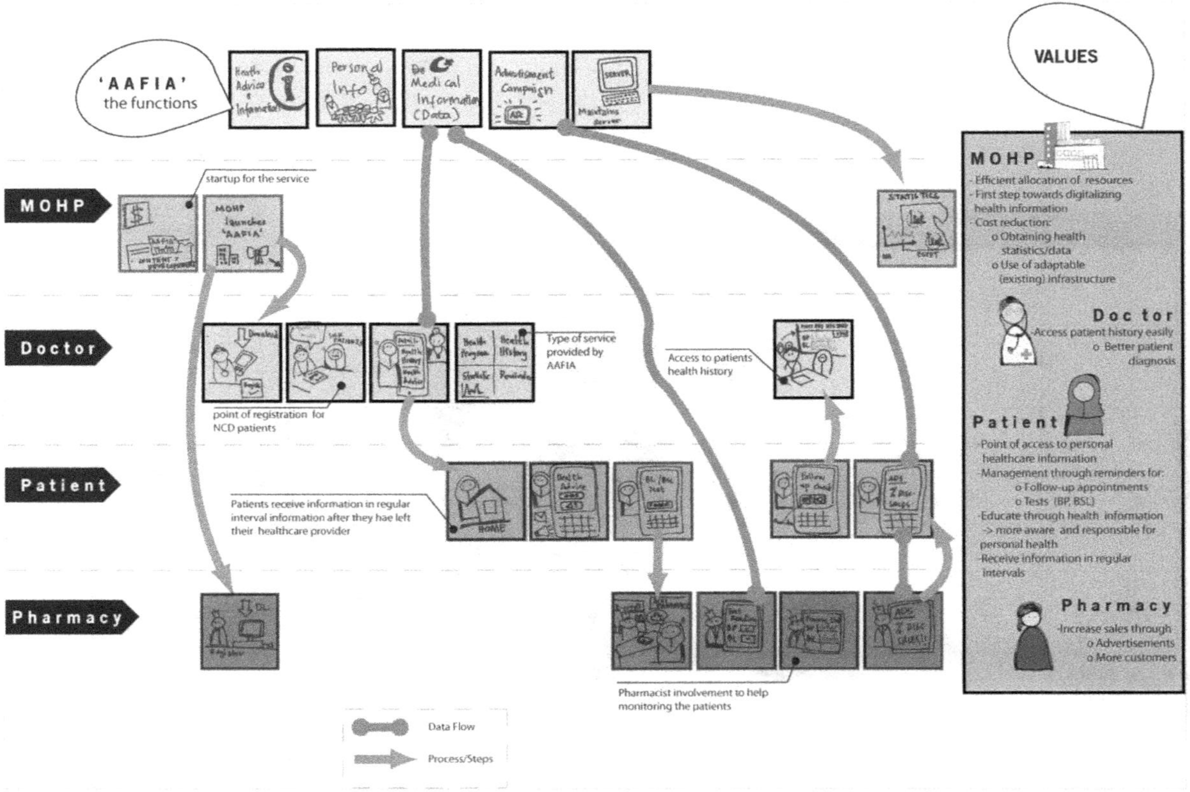

FIGURE 9.11   Data flows and process step dependencies among stakeholder processes.

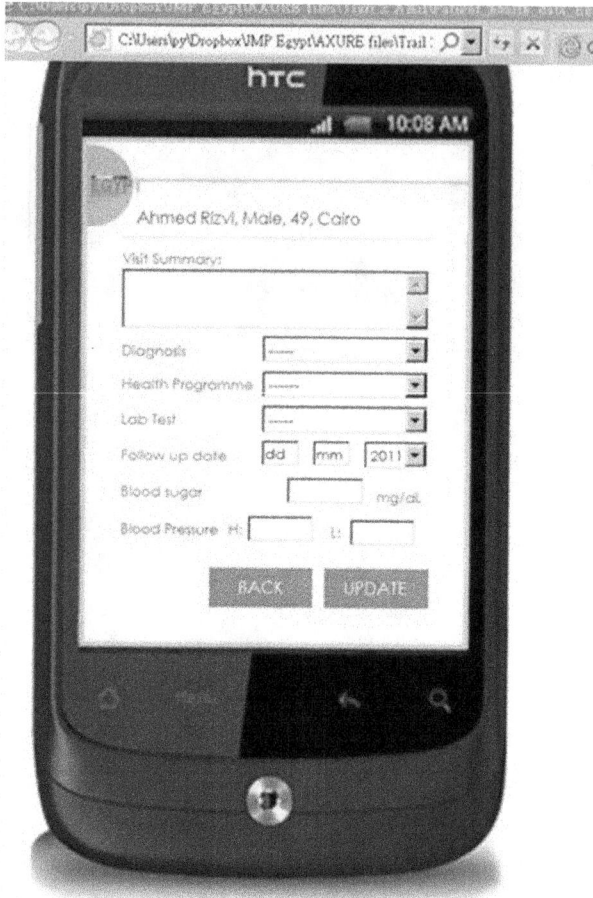

FIGURE 9.12   Axure RP prototype.

to them. This would give them direct and easy access to information and could be a more efficient way to transfer messages to patients in rural areas.

- Private doctors will get some basic information about their patients, as well as a platform to keep data to monitor their NCD patients. As the system expands, doctors should also be able to add to the health database, including their personal advice, by paying a small fee, which can be a way to promote their services within this competitive market. Figure 9.13 depicts these potentialities.

Future Vision

This system provides long-term benefits as it reduces costs incurred by healthcare providers and patients, especially for tertiary care, by trying to shift the focus towards better primary healthcare. Since the system supports the monitoring and prevention phases, it reduces patients' risk of moving into serious stages of illness or prolonging their current adverse medical state. If the currently designed core system functions

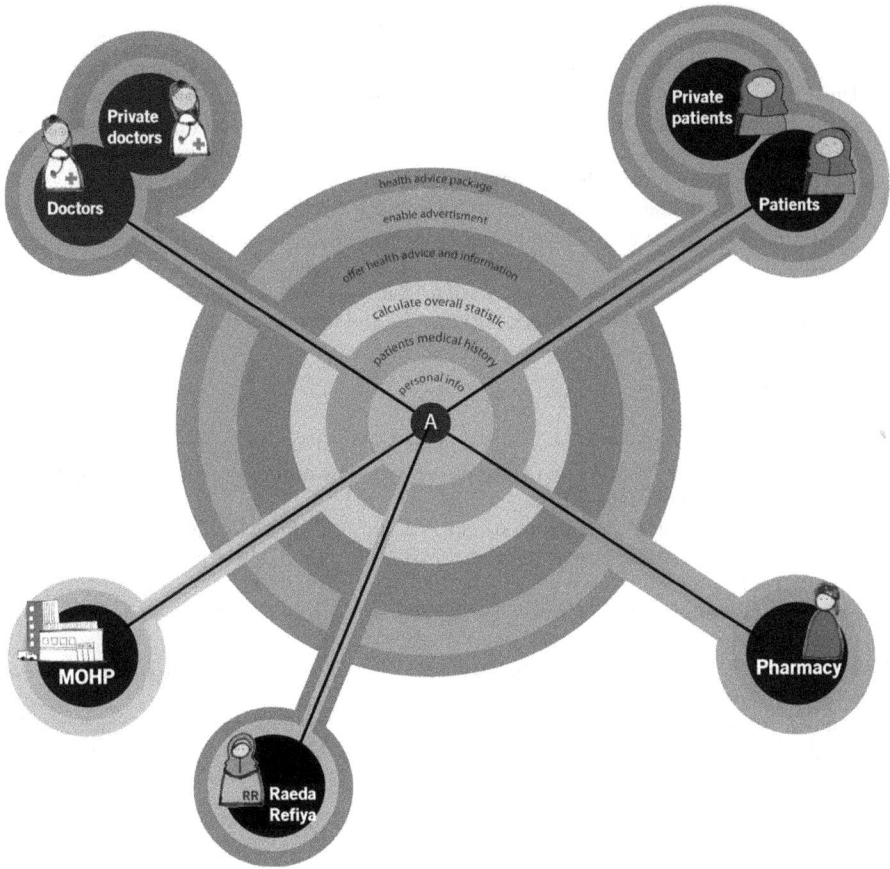

FIGURE 9.13 Potential future system expansions.

well, it can then be expanded towards more profitable services that generate revenue, for example, through paid users as in the case of private doctors and their patients.

CONCLUSIONS AND LESSONS LEARNED

Designing a healthcare intervention for use by the public, many of whom may be illiterate or poorly educated, is a complex undertaking. This project demonstrated that a culturally aware, technologically accessible solution can improve NCD management in low-resource settings. Key lessons include:

- Design tools must reflect user context, particularly literacy and cultural norms.
- All stakeholders must perceive value from the system. Key stakeholders must collaborate to "lift" the intervention and make it operationally successful and economically viable.
- Scalable, sustainable implementation depends on ecosystem-wide integration.

NOTES

1 This design initiative was couched within a joint MSc project by students from Cairo University and Technical University (TU) Delft, supervised under the EU Erasmus program funded UNCHAIN (University Chair on Innovation) project, the University Chair on Innovation concept itself was originally proposed by UNIDO. Respective University Chairs on Innovation from the two universities shared the directing of the project.
2 The team is particularly grateful to Prof. Medhat El-Refaie, Professor of Cardiology at Cairo University School of Medicine, and Head of the Egyptian Curative Organisation—at the time of this project—for instigating this project with his idea and insight.
3 The workspace for the research and design team was very kindly loaned by Professor Aboulela Hassanein, of the Faculty of Computers and Information, Cairo University.

REFERENCES

Central Agency for Public Mobilization and Statistics (CAPMAS) (2025a). https://www.sis.gov.eg/Story/197754/CAPMAS-Egypt's-population-officially-reaches-107mn?lang=en-us. Accessed 6th June 2025.

Central Agency for Public Mobilization and Statistics (CAPMAS) (2025b). https://www.capmas.gov.eg/Admin/News/PressRelease/202198111315_999-e.pdf. Accessed 8th June 2025

Fasseeh et al. *Journal of the Egyptian Public Health Association* (2022) 97:1. https://doi.org/10.1186/s42506-021-00089-8

Hofstede, G. (2011). Dimensionalizing cultures: The Hofstede model in context. *Online Readings in Psychology and Culture*, 2(1):8.

Radwan G, Adawy A. The Egyptian health map: a guide for evidence-based decision making. *East Mediterr Health J.* 2019 July 24;25(5):350–361. https://doi.org/10.26719/emhj.18.048. PMID: 31364760.

Section 3

Enriching Design Practice
through Multidisciplinary
Collaboration and Integration

10 The Evolving Landscape of UX Design

Balancing Empathy and Efficiency

Rama Vennelakanti and Joshua Ekandem

BACKGROUND

THE RISE OF UX: UBIQUITOUS COMPUTING, ENGINEERING CULTURE, AND THE SPRINT MENTALITY

The rise of ubiquitous computing and consumer electronics has made User Experience (UX) a key differentiator in the era of smartphones and digital interfaces (Harvey et al., 2022). UX originally emerged from the need of engineers to address errors of untrained users operating complex machinery (Vinney, 2023). Today, smartphones are integral to the daily life of over 3.5 billion people around the world (Gu, n.d.). In response, companies have invested heavily in User Interface (UI) design to ensure usability, accessibility, and brand coherence across platforms, and UX designers are expected to adapt and keep pace with fast-paced updates.

The advances in software development, particularly the proliferation of Agile methodologies, have significantly impacted UX practices. The processes aimed at shortening development cycles have also compressed research and design timelines, with delivery speed prioritized over user insight (Dayton & Barnum, 2009). In this sprint-driven environment, UX work has become increasingly reactive—focused on shipping quick fixes rather than addressing systemic user needs. Unfortunately, this often results in the implementation of minimally viable solutions rather than truly holistic, user-centred ones.

SITUATING THE UX TEAM AND ITS PLACE IN THE HIERARCHY

While Bauhaus embraced industrial production and even celebrated the industrialization undertaken by Ford, its core goal was to democratize good design, making well-crafted, high-quality products accessible to all, rather than serving only elite or profit-driven interests (Contrary, n.d.; Gropius House, n.d.). However, in practice, its ideals were later appropriated by capitalist systems, leading to minimalist aesthetics without the context or refinement (Heller, 2024). For example, the commodification of corporate modernism with everyone trying to be different and modern,

DOI: 10.1201/9781003557777-13

but everyone started to look alike with more cheaply and less skilfully built reproductions (Architizer, n.d.). We are up against a tension, or should we say a dance, between how UX is practiced and why it is practiced. Forcing practitioners to define if the goal is of UX efficiency and ease, or cultural and emotional relevance?

A notable challenge in defining the goal is that UX practitioners often find themselves caught between business objectives and user needs, struggling to balance corporate priorities with genuine user advocacy. Instead of being facilitators, they are frequently forced to justify their role, navigate internal power struggles, and adapt to an environment where efficiency and automation increasingly overshadow thoughtful, user-centred design.

Where the UX team sits often determines the direction and impact it has (Delve, n.d.). A simplistic view of this situation puts the UX practitioner at the centre, but we have seen this tension since the early 20th century. In some organizations, UX is treated as an independent function on par with product, engineering, marketing, etc.; in other organizations UX is attached as a sub-function reporting to the product, engineering, or marketing. In the latter cases, a UX team may be constantly required to defend their value beyond short-term solutions. Each situation gives the UX team an edge or constrains it. As Figure 10.1 illustrates:

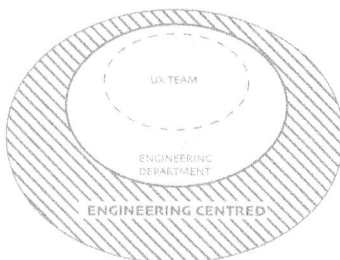

UX reports into to engineering

+ closely aligned to the engineering roadmap
+ quick to respond to engineering requirements
+ aligned to the platform capabilities and constraints
- mostly driven by tactical requirements
- design is often limited to wireframes and buttons

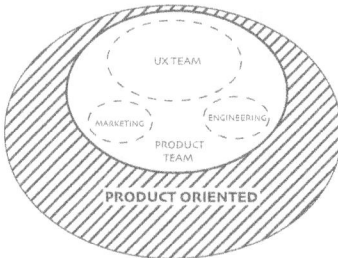

UX is a part of the product team

+ closely aligned with the product roadmap
+ usually driven by quantifiable metrics
- likely to consider market research as sufficient and a replacement of UX research
- likely to consider market imperatives over user requirements

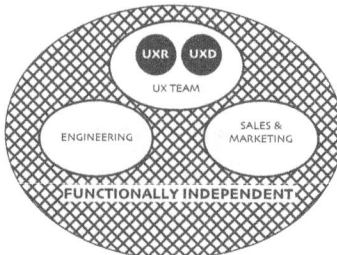

UX is an independent function

+ often drives strategic research
+ can ensure synergy with organizational roadmaps
- can be considered an outsider; not with it in the trenches
- can be treated as a roadblock, or time consuming luxury

FIGURE 10.1   Organization structures. Illustration created by the author.

THE ERA OF UX COMMODIFICATION: PITFALLS OF MODERN-DAY UX

TOOL-DRIVEN TRAINING

The UX workforce has shifted focus, towards mastering tools like Figma for UI design and ABTasty for A/B testing, from understanding the foundational design-research principles behind deep, user-centred design. Proficiency in these tools has become an essential credential on résumés to land a job. Bootcamps and fast-track programmes offer quick certifications, emphasizing marketable skills over deep expertise, and creating a pipeline of designers trained to execute rather than question, reducing UX to a checklist of tasks. As expertise has been bundled into a single role (e.g. UX/UI) and scaled rapidly, the profession has moved away from its artisanal roots. One very public example of this shift is with Jony Ive, Apple's former Chief Design Officer until 2019, who championed minimalism, utility, and meticulous attention to detail. Ive's work at Apple, including the iMac (1998), iPod (2001), iPhone (2007), and Apple Watch (2015), transformed the company into a cultural phenomenon. His departure in 2019 marked a shift in the industry, where design began to take a backseat to efficiency and functionality (Feiner, 2019). The demand for "prima-donna" designers, who prioritized perfection, has given way to formulaic approaches focused on efficiency and replicability. This commodification has allowed organizations to scale design teams quickly, however, at the cost of depth, nuance, and a true understanding of the UX.

THE DESIGNER'S ERODING INFLUENCE AND THE RESEARCHER'S GROWING FRUSTRATION

With depth being deprioritized in design education, these newly minted designers often lack the broader perspective to navigate cross-functional collaboration or defend their role meaningfully. Far from the romanticized image of a designer as a visionary creative, designers today often feel sidelined, reduced to styling screens rather than shaping vision. Once influential creators, many now wish for an MBA or engineering degree to better arm themselves to have a seat and a voice of influence in strategic decisions. Being forced to follow technical constraints and trend cycles, with limited strategic input, leads to an insular and defensive culture. Instead of welcoming critique and discussion, many online design forums reflect just "show and tell" of a designer's work. Domains once complementary are now viewed as encroachments, leading to tension with other stakeholders and raising the question: has design become too guarded to grow?

Similarly, far from the essential view of informative truth-tellers, today's UX researchers often feel not heard. Once valued for their insights into user behaviour and needs, many now struggle to prove their value in rapid product cycles. The challenges of bridging the gap between strategic research and the fast-paced demands of modern product development force researchers to be heads down in their silo just trying to maintain the flow of work coming in to keep up. Researchers are left producing reports that gather dust or get cherry-picked for data that supports predetermined decisions. This shift has reduced researchers to data gatherers, their nuanced findings often simplified or ignored in favour of quick, actionable metrics. Valued less for strategy and more for speed, many researchers find themselves in reactive and defensive stances. Online UX research communities reflect this frustration, with discussions often centred on how to prove the worth of research rather than advancing methodologies. This defensive posture raises a critical question: has UX research become too isolated from the product development process, struggling to demonstrate its value?

COLLABORATION VS. OWNERSHIP TENSIONS

UX is impacted and impacts every stage of the user's interaction with the product or service and must effectively collaborate to deliver a product (experience) (Interaction Design Foundation, n.d.). At every stage of defining, building, and delivering the UX, the notional lead shifts between the critical stakeholders—research, design, and engineering. This transitioning of responsibility to lead and deliver, in turn, impacts how the teams collaborate. In practice, this is a process fraught with tensions as the ownership of the product often rests outside of the UX team, with the product owner. UX requires collaboration across disciplines, but sometimes the need for shared ownership leads to conflict. Teams compete to "own" the experience, undermining the complementary strengths each brings. True UX success lies in applying the right expertise at the right time, not in dominance.

THE NEED FOR A PARADIGM SHIFT

This trajectory has brought us to a crossroads. As UX has become increasingly entangled with engineering and technological priorities, its potential to drive meaningful change has been sidelined. Instead of a balanced approach that combines empathy with efficient practices, this emphasis on tools and metrics has shifted focus from empathy and user understanding, resulting in misguided products and services that compromise the quality of the same experiences they were supposed to improve. Moving forward, designers must reclaim their role as strategic partners, using emerging technologies not to replace creativity and empathy but to amplify them. This realignment is essential for better product and service outcomes in the face of evolving technological and social demands. The profession must reckon with this shift and refocus on its original purpose: understanding and advocating for users.

UX is about designing experiences that help users reach their goals. Goals that often reveal themselves through moments of friction or flow. Whether observed in the wild or under controlled conditions, the many paths users take to navigate to their goals are as textured and intricate as baskets crafted by the Efik people of Nigeria. Similarly, UX researchers and designers thrive on a rich tapestry of evolving methods that must be developed and applied relevantly for their specific context and goals. Early-stage methods such as exploratory research, generative techniques, co-design, and feasibility assessments help lay the groundwork for later evaluative approaches like A/B testing. This sequence enables more informed, strategic decision-making. The coming together of the different disciplines requires proactive informing and being informed by the other. This call for this synergistic performance is aimed at ensuring that we do not lose sight of who this is for—*The User*!

Spotlight User Needs and Contexts

At the heart of this reform is a renewed focus on addressing genuine user needs. While technology and platforms are important, they should be viewed as constraints (the rules for the space) rather than boundaries (the space itself). UX practitioners must reaffirm that the real users are not those designing or building the product, but a diverse group with varying capacities and capabilities, including both intended and unintended users. Understanding context is crucial in this refocusing effort. The user's context extends far beyond design, technology platforms, and digital real estate, encompassing real-world environments such as shop floors, farmlands, hospitals, and classrooms. These rich, diverse settings should inform design decisions.

At the heart of value-based UX is a return to understanding users in their contexts and their evolving needs as they experience and interact with contexts, products, and services towards their goals. This means prioritizing localized, contextual, and co-design approaches to address current gaps in understanding user needs. Realize that in the digital world, often the world we live in today, the context of use that real design emerged out of is often absent. Observation is the foundation for innovation and relevance, yet it is a skill that is often undervalued in contemporary UX practices.

By centring design on observation and empathy, the profession can reclaim its broader scope and have a more impactful scope. What would it look like to train designers differently? Imagine prioritizing UX researchers with exceptional communication skills over the traditional "rockstar" designer. The design process itself—iterative, structured, and teachable—can be mapped and refined. But the ability to uncover human stories and distil them into actionable insights is what drives meaningful solutions.

Develop Transferable Skills and Interdisciplinary Communication

In the dynamic, tech-driven world of UX, developing transferable skills is essential to adapt and grow. Practitioners should focus on building sound foundations in design and behavioural sciences, while using bootcamps and tutorials to upskill. This approach helps professionals avoid being displaced by the automation that platforms increasingly offer. Furthermore, effective communication with adjacent disciplines is crucial. Understanding how these disciplines interact with and inform each other allows UX professionals to better integrate diverse perspectives and create more holistic solutions. By embracing these reforms, the UX discipline can reclaim its pivotal role in creating meaningful, user-centred experiences. This renewed focus on user needs, context, diverse methodologies, and interdisciplinary skills will empower UX professionals to navigate the complexities of modern technology landscapes while staying true to the core principles of user-centred design.

Co-Opt AI as a Tool and Aid

We believe in the immense value that UX research and design bring to both the industry and society. This value is rooted in the vast, yet often underutilized, collective knowledge and insights that the UX community has developed over time. While artificial intelligence (AI)-powered by large language models can process vast amounts of secondary textual and speech-to-text data, it still cannot grasp the non-verbal, in-situ subtleties of human behaviour that are central to understanding an experience. These nuances—visual, auditory, contextual—remain the domain of human researchers and designers. The richness of lived experience still eludes automation.

However, AI can expand capabilities by reducing repetitive work and enriching knowledge repositories. AI can assist in maintaining consistency, automating UI component generation, and even drafting business artifacts within minutes. As a result, the role of UX designers is shifting. It is no longer enough to simply craft interfaces; designers must become experience architects, tackling deeper systemic challenges that shape how people interact with technology—and with each other.

Advances in AI and automation have reshaped the design landscape. Today, one person equipped with the right tools can achieve what once required entire teams. But rather than threatening the relevance of UX professionals, these technologies provide an unprecedented opportunity: to free designers from routine tasks and focus their efforts on solving complex, global problems.

The UX designers of tomorrow must leverage these tools to abstract the mechanics of design and shift their attention to issues like accessibility, sustainability, and equity. Designers empowered by AI can turn their expertise towards improving societal systems—reimagining services, creating communities, and helping people live better lives together. By integrating these capabilities into design practice, we can align UX with the challenges of our time. The result is a discipline capable of crafting experiences that are not just usable but transformative, meeting the needs of both individuals and the world they inhabit.

DEFINE NEW PARADIGMS AND FRAMEWORKS

In conclusion, with all the transformative advances in the field of technology and their impact on people and the fields of design and research, there is a need to realign and craft new strategies to stay relevant and have an impact. Here are some ways new frameworks can account for these shifts:

- Develop frameworks for design with AI, and for continuous integration; like Bauhaus embraced the process of industrialization but held onto its ideals.
- Integrate AI as an aid by defining best practices to use it to amplify human effort and not just replace it.
- Highlight and cultivate synergies with adjacent disciplines to understand how to complement each other.
- Emphasize skills and knowledge required to collaborate with functions and workflows that contribute to and implement design.

REFERENCES

Architizer. (n.d.). *New corporate modernisms*. Architizer. Retrieved April 16, 2025, from https://architizer.com/blog/inspiration/collections/new-corporate-moderisms/#:~:text=Corporate%20modernism%20is%20a%20term,half%20of%20the%2020th%20century

Contrary. (n.d.). *Industrial designers*. Contrary. Retrieved April 16, 2025, from https://www.contrary.com/blog/industrial-designers

Dayton, D., & Barnum, C. (2009). The impact of Agile on user-centered design. *Technical Communication*, *56*(3), 219–234.

Delve. (n.d.). *Interaction design beyond the screen*. Delve. Retrieved April 16, 2025, from https://www.delve.com/insights/interaction-design-beyond-the-screen

Feiner, L. (2019, June 28). *Jony Ive's departure marks the end of the hardware era at Apple*. CNBC. Retrieved April 16, 2025 https://www.cnbc.com/2019/06/28/jony-ives-departure-marks-the-end-of-the-hardware-era-at-apple.html

Gropius House. (n.d.). *Bauhaus manifesto*. Gropius House. Retrieved April 16, 2025, from https://gropius.house/location/bauhaus-manifesto/

Gu, Tianyi (n.d.). *43% of Active Smartphones Will Be 5G-Ready by 2023: The Global Mobile Market Is on Track for Substantial...Newzoo*. Retrieved April 16, 2025 from https://newzoo.com/resources/blog/mobile-game-market-2020-smartphone-users-game-revenues-5g-ready-engagement#:~:text=The%20Mobile%20Market's%20Unstoppable%20Growth,of%20smartphone%20users%20in%202020

Harvey, D. L., Milton, K., Jones, A. P., & Atkin, A. J. (2022). International trends in screen-based behaviours from 2012 to 2019. *Preventive Medicine*, *154*, 106909.

Heller, S. (2024, May 22). *The Daily Heller: Eliot Noyes, Godfather of Corporate Modernism—The Movie*. PrintMag. https://www.printmag.com/daily-heller/the-daily-heller-eliot-noyes-godfather-of-corporate-modernism-the-movie/

Interaction Design Foundation. (n.d.). *What is user experience (UX) design?*. Interaction Design Foundation Retrieved April 16, 2025, https://www.interaction-design.org/literature/topics/ux-design

Vinney, C. (2023, January 9). *The complete history of UX (User experience)*. UX Design Institute. Retrieved April 16, 2025 from https://www.uxdesigninstitute.com/blog/history-of-ux/

11 Engineering Enjoyment
Empirical Research on Designing Fun Interactive Experiences

Owen Schaffer

INTRODUCTION

A player sits at their computer at 2 AM, having spent six hours navigating challenging puzzles and obstacles. Rather than feeling frustrated or fatigued, they remain energized and completely focused on the task at hand. Hours have passed unnoticed. Elsewhere, a student is playing an educational game, excited to solve the next problem because the design makes learning feel like play. These moments of deep engagement happen when designers understand what draws people in and keeps them motivated.

Creating interactive systems that people will enjoy using and will want to use matters in many different fields. Game developers want to build experiences that players love and return to again and again. Designers of websites and apps try to make their interfaces enjoyable to use, not just functional. Educational software developers work to make learning feel more like an adventure than a chore. Healthcare game developers look for ways to use interactive systems to help people stick with healthy behaviors. At the center of these efforts is the same underlying question: What makes people enjoy using interactive systems?

This chapter presents methods and tools for understanding and designing enjoyable interactive experiences based on empirical research about what makes digital games enjoyable. Digital games offer an ideal starting point because they are designed specifically for player enjoyment. Digital games are interactive, computer-based systems that offer goal-directed, challenging tasks designed for enjoyment. Enjoyment refers to how positively people evaluate their experience. The sources of digital game enjoyment are the specific positive experiences that lead to this positive evaluation.

The implications of this research extend well beyond gaming. The same psychological principles that make games compelling can inform the design of educational software, productivity applications, health interventions, and any interactive system where user engagement matters. When we understand what creates intrinsically rewarding experiences, we can apply these insights through gamification, incorporating game design elements into non-game contexts, and serious games, games for purposes beyond entertainment (Deterding et al., 2011).

DOI: 10.1201/9781003557777-14

By grounding interactive design in empirical research on enjoyment, creators can move beyond guesswork and intuition to develop experiences that systematically fulfill users' psychological needs and desires. Just as positive psychology shifted focus from treating mental illness to increasing human flourishing, this positive user experience approach moves beyond fixing usability problems to creating fulfilling experiences that satisfy needs and desires and increase user well-being, happiness, or quality of life. The frameworks and methods presented in this chapter provide practitioners with evidence-based tools to design digital experiences that users find intrinsically rewarding, satisfying, and meaningful.

The remainder of this chapter unfolds in six main sections, each building toward a more comprehensive understanding of how to design for enjoyment. Section "Theoretical Foundations of Digital Game Enjoyment" establishes the theoretical foundations by reviewing diverse academic literature spanning psychology, game design, anthropology, and human-computer interaction (HCI) to identify the many factors that contribute to positive interactive experiences. Section "The Critical Role of Flow in Enjoyable Experiences" examines the critical role of flow theory in understanding enjoyable experiences, introducing the Feedback Loop of Flow model, which explains the process of cognition and action people go through when their attention is fully focused on an enjoyable, intrinsically motivating activity and the key system design elements needed for users to get into that focused, enjoyable flow state. Section "Empirical Investigations of Digital Game Enjoyment" presents empirical investigations of the sources of digital game enjoyment, detailing a card sorting study with game players to identify sources of enjoyment and the development of the *Enjoyment Questionnaire (EQ)* and *Sources of Enjoyment Questionnaire (SoEQ)* that measure enjoyment and 38 sources of enjoyment. Section "Evaluation Methods for Player Experience" outlines evaluation methods for player experience research, including both applied research approaches for improving specific games and basic research methods for generating generalizable knowledge about enjoyment and its sources. Section "Applications of Player Experience Research" explores practical applications of this research in game design, serious games, and gamification, while addressing the ethical considerations that arise when designing systems intended to be enjoyable, engaging, and motivating. Finally, Section "Conclusion and Future Directions" provides conclusions and identifies future research directions for advancing our understanding of enjoyable interactive experiences.

THEORETICAL FOUNDATIONS OF DIGITAL GAME ENJOYMENT

Understanding what makes people enjoy playing digital games is important for designing interactive systems that people want to use. This knowledge is important not only for game developers but also for HCI researchers who want to create systems that people will find intrinsically motivating and enjoyable to use. To design for user engagement and retention, we need to understand what creates positive experiences when people play digital games.

Digital game enjoyment is how positively people evaluate their experience while playing a digital game. The specific experiences that lead to this positive evaluation are defined here as the sources of digital game enjoyment.

Positive psychology studies positive human traits, experiences, relationships, and institutions (Seligman & Csikszentmihalyi, 2000). The field produced a classification of 24 Character Strengths and Virtues (CSV) that generate unique positive experiences when practiced (Park et al., 2004; Peterson & Seligman, 2004). The CSV was intended to be positive psychology's counterpart to clinical psychology's *Diagnostic and Statistical Manual of Mental Disorders* (DSM; American Psychiatric Association, 2013).

Positive psychology research has identified three paths to happiness: flow, pleasure, and meaning (Peterson et al., 2005). The pleasure pathway emphasizes maximizing sensory pleasure while minimizing pain and discomfort. The meaning pathway involves taking action driven by prosocial purposes such as helping others, improving society, or supporting a cause. The flow pathway involves overcoming challenges for the intrinsic enjoyment that the experience of overcoming the challenges provides.

Positive psychology research on positive emotions has identified many emotional states with specific evolutionary functions. Fredrickson's work identified 10 positive emotions, including serenity, interest, hope, pride, and inspiration (Fredrickson, 2009). Shiota's taxonomy of positive emotions examined the adaptive benefits of eight emotions, including enthusiasm, contentment, nurturant love, amusement, and awe (Shiota, 2014). Their work explored how each positive emotion may have served survival and reproductive advantages throughout human evolution (Shiota et al., 2014). Other research from Condon and colleagues has suggested that emotions typically experienced as negative can have pleasant forms (Condon et al., 2014). Examples include the fear experienced on roller coasters, the anger directed at game villains, or the sadness experienced during poignant narratives.

Self-Determination Theory proposes that fulfilling three psychological needs fosters intrinsic motivation and enjoyment: competence, relatedness, and autonomy (Deci & Ryan, 1985; Ryan & Deci, 2000). The Player Experience of Need Satisfaction model extended this model to the context of player motivations to play games, incorporating the original three needs while adding presence and intuitive controls (Ryan et al., 2006). Presence refers to the feeling that one is there in the game world. Intuitive controls refers to how easy it is for players to learn and use the controls of the game.

Reiss and Havercamp's motivational theory proposed 16 human desires that drive behavior (Havercamp, 1998; Reiss, 2002, 2004; Reiss & Havercamp, 1998). These desires are power, curiosity, independence, status, social contact, vengeance, honor, idealism, physical exercise, romance, family, order, eating, acceptance, tranquility, and saving. Measures of these desires have been adapted for studying digital game enjoyment (Schaffer, 2019), and each of these desires may represent a potential source of digital game enjoyment.

Analyses of play may offer insights into its cultural and psychological functions. Sutton-Smith's examination identified multiple potential sources of enjoyment, including advancement, chance, dominance over others, identity formation, imagination, peak experiences, and pure frivolity (Sutton-Smith, 2009). His work explored how different cultures develop rhetorics of play reflecting their values about the purposes and benefits of play. Caillois's classification distinguished four game types: competition (agon), chance (alea), simulation and role-playing (mimicry), and vertigo-inducing disorientation (ilinx) (Caillois, 1961). Each of these categories was a potential source of enjoyment.

Frameworks have emerged for understanding digital game engagement. Player typology research began with Bartle's identification of four player orientations based on preferred virtual world activities (Bartle, 1996). These are Achievers who focus on goal completion, Socializers who seek interpersonal connection, Killers who pursue player-versus-player dominance, and Explorers who emphasize world discovery. Yee's empirical work refined these concepts into three motivation components: achievement, social interaction, and immersion (Yee, 2006; Yee et al., 2012). More recently, Yee identified six clusters of gaming motivations: Action (Excitement & Destruction), Social (Collaboration & Competition), Mastery (Strategy & Challenge), Achievement (Power & Completion), Creativity (Design & Discovery), and Immersion (Story & Fantasy) (Yee, 2016).

Play personality research examined how individual characteristics influence game preferences. Brown and Vaughan identified eight play personalities: The Joker, Kinesthete, Explorer, Competitor, Director, Collector, Artist/Creator, and Storyteller (Brown & Vaughan, 2010). Fullerton expanded this framework by adding The Achiever and The Performer, suggesting that each personality type may derive enjoyment from particular gameplay activities (Fullerton, 2014). The activities of these player types may each be sources of digital game enjoyment.

Lazzaro's "four keys to fun" model categorized pathways to digital game enjoyment (Lazzaro, 2004, 2009). These pathways are Easy Fun (driven by novelty and curiosity), Hard Fun (centered on challenge and triumph), People Fun (focused on friendship and amusement), and Serious Fun (involving altered consciousness and relaxation). Her research also identified more emotions game players experience, including Fear, Surprise, Naches, Fiero, and Schadenfreude (Lazzaro, 2004). Naches is a Yiddish term for satisfaction from others' achievements. Fiero is an Italian term for triumph and pride experienced after succeeding at a challenging task. Schadenfreude is a German term for pleasure from others' misfortune.

The Playful Experiences (PLEX) framework synthesized previous models into 20 categories of playful experiences (Korhonen et al., 2009). These categories include Completion, Discovery, Relaxation, Sensation, Expression, Subversion, and Fellowship. This framework drew on previous models, particularly Costello and Edmonds's pleasure framework (Costello & Edmonds, 2007). The framework was developed into a set of design cards used in game development projects (Lucero & Arrasvuori, 2010, 2013. It has also been developed into a Playful Experiences Questionnaire (*PLEXQ*) with 17 playful experiences clustering into four factors: stimulation, pragmatic, momentary, and negative experiences (Boberg et al., 2015).

Empirical studies have investigated the sources of digital game enjoyment using quantitative methods. Quick et al. used factor analysis to identify six dimensions of game enjoyment: Fantasy, Exploration, Fidelity, Companionship, Challenge, and Competition (Quick et al., 2012). Abeele et al. developed the *Player Experience Inventory*, measuring 11 aspects of player experience (Abeele et al., 2020). These aspects are meaning, mastery, immersion, autonomy, curiosity, ease of control, challenge, progress feedback, audiovisual appeal, goals, and rules. Their work demonstrated how these factors cluster into two high-level factors, functional and psychosocial consequences, that impact overall enjoyment. While such higher-order categorizations may provide valuable theoretical organization, examining sources of enjoyment at a more granular level may offer additional insights into which specific mechanisms most contribute to positive player experiences.

THE CRITICAL ROLE OF FLOW IN ENJOYABLE EXPERIENCES

Flow theory is central to understanding enjoyment. Flow is the experience of overcoming challenges for the intrinsic enjoyment derived from tackling those challenges while continuously adjusting performance based on feedback. Flow is the psychological experience of effortless concentration of attention on an enjoyable and interesting activity one wishes to be doing (Csikszentmihalyi & Nakamura, 2010; Hektner et al., 2007).

FLOW CONDITIONS AND INDICATORS

It is important to distinguish between the conditions that lead to or cause flow and the indicators that measure how much a person is in a flow state. While much research on flow has attempted to measure it using questionnaires based on all nine dimensions identified by Csikszentmihalyi (Csikszentmihalyi, 2008; Csıkszentmihályi, 1975), such as the measure of flow in computer games developed by Fang and colleagues and the *Flow State Scale* developed by Jackson and colleagues (Fang et al., 2013; Jackson & Eklund, 2004; Jackson & Marsh, 1996), a more precise approach involves separating the flow conditions that flow theory suggests cause flow from the flow indicators that assess how much a person is experiencing a flow state.

Flow theory proposes that three flow conditions are necessary and sufficient to get people into a flow state: optimal challenges that stretch skills, clear proximal goals, and immediate progress feedback. These conditions were discovered through qualitative interview research by Csikszentmihalyi (Csikszentmihalyi, 1993, 2008; Csikszentmihalyi & Csikszentmihalyi, 1988; Csıkszentmihályi, 1975) and further articulated by Nakamura and Csikszentmihalyi (Nakamura & Csikszentmihalyi, 2014).

Optimal challenges is how much the person feels that a task is challenging but feels their skills are sufficient to meet those demands, meaning perceived challenges and perceived skills are balanced and high. Perceived challenges and skills need to be high to stretch the person's skills. The original 3-channel model of flow suggested that when challenge is too low for skills, a person would feel bored, and when challenges are too high for skills, a person would feel overwhelmed and anxious (Csıkszentmihályi, 1975). Clear proximal goals refer to how much the person feels the objective is clear for each immediate next step in the activity, rather than just the clarity of the overall goal. As explained by Csikszentmihalyi and Nakamura, the focus is on knowing what to do at each moment in the action sequence (Csikszentmihalyi & Nakamura, 2010). Immediate progress feedback is how much the person feels they know how well they are performing the activity and making progress toward their goals, allowing them to adjust their actions to improve their performance (Csikszentmihalyi et al., 2014).

Flow indicators are the factors that indicate how much a person is in a flow state. Building on Csikszentmihalyi's nine elements of flow (Csikszentmihalyi, 1993,

1998, 2008), and excluding the three conditions already discussed, the remaining indicators include concentration, sense of control, merging of action and awareness, loss of self-consciousness, altered perceptions of time, and autotelic experience. Hektner et al. suggested that flow can be measured through a combination of concentration, enjoyment, and either interest, desire to be doing the activity, or excitement (Hektner et al., 2007).

TEST-OPERATE-TEST-EXIT (TOTE) UNITS THEORY

Test-Operate-Test-Exit (TOTE) units theory, developed by early cognitive psychologists, shows how feedback guides people toward goals. Given that flow is experienced during engagement in goal-directed tasks, this theory helps explain the perceptual, cognitive, and action steps that generate the flow experience.

TOTE units were proposed by Miller et al. as an alternative framework to Skinner's stimulus-response pathways (Miller et al., 1960). Instead of viewing behavior as a reaction to external stimuli, TOTE units describe a feedback loop. The test phase involves comparing the current state with the desired goal. If the goal has not been met yet, the person continues to the operate phase, taking action to achieve the goal. After operating, the person tests again to see if the goal has been met. This cycle of test, operate, and test repeats until the test indicates that the goal has been achieved, at which point the person exits the activity. Hammering a nail flush is a classic example of a TOTE unit in action, testing to see if the nail is flush with the wood while hitting the nail with a hammer until the test shows that the nail is flush and the goal has been achieved.

TOTE units are recursive and can exist at multiple levels of abstraction or specificity simultaneously. The operate phase of a TOTE unit can itself consist of a sequence of TOTE units or atomic operations, forming hierarchical plans for behavior. This nested structure allows for complex, goal-directed activities to be organized into a hierarchy of TOTE unit feedback loops. The fundamental feedback loop within TOTE units is considered a building block of the cognitive process necessary for people to accomplish goal-directed tasks. Considering that flow experiences also occur during goal-directed tasks and involve continuous adjustment of actions based on feedback, -TOTE units theory is well-suited as a foundation for a framework to understand the process of cognition, perception, and action that generates the flow state.

THE FEEDBACK LOOP OF FLOW MODEL

Schaffer and Fang proposed a feedback loop of flow model for computer-based tasks, which builds on the foundation of Test-Operate-Test-Exit (TOTE) units theory (Schaffer & Fang, 2022). This model integrates the process described by TOTE units with the three flow conditions identified in flow theory to model the steps of cognition, perception, and action that generate a flow state. While TOTE units include the steps of taking action and testing progress toward a goal, they do not explicitly address the initial step of setting the goal that is tested in the testing phase (Miller et al., 1960). To fill this gap in the model and incorporate the clear proximal goals flow condition, the feedback loop of flow model introduces a goal-setting step in the loop before testing, reflecting the cognitive process of choosing the desired outcome against which progress is tested in the testing progress step. This model also shows how specific design elements of computer systems can support each of the three flow conditions.

Figure 11.1 presents a simplified version of the Feedback Loop of Flow to clarify how the three flow conditions, which are also sources of enjoyment (see Table 11.1), provide the necessary and sufficient inputs for the different parts of the feedback loop process. The figure combines and simplifies two models from Schaffer and Fang's work: the feedback loop of flow and the design for flow models (Schaffer & Fang, 2022). The original models distinguished between system design elements (features that can be changed and differ between systems) and flow conditions (users' perceptions of how well these conditions are being met, measurable through surveys). This simplified version focuses on the flow conditions and how they fuel the cognitive feedback loop process.

In a flow state, people engage in a recurring sequence of setting proximal goals, checking progress, taking action, and testing their progress again, continuing this feedback loop until the goal is either achieved or abandoned. The goal-setting step is an important step forward from TOTE units theory. To return to the example of hammering a nail, a person must first set the goal of hammering the nail flush before swinging the hammer to have a specific outcome to test in the testing phase. People can also adjust their goals based on the feedback they receive. If the nail will not go in any further, they may change their goal to removing the nail to try again, or leaving it as it is.

Similar to TOTE units (Miller et al., 1960), the feedback loop of flow operates recursively at different levels of task abstraction or granularity. The action step can itself consist of subtask feedback loops with the same structure, continuing down to

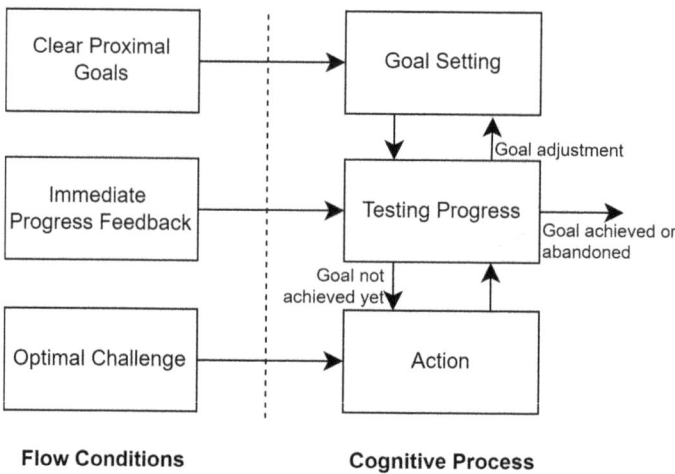

FIGURE 11.1   The Feedback Loop of Flow, adapted from Schaffer and Fang (2022).

TABLE 11.1
Factors of the Sources of Enjoyment Questionnaire (Schaffer, 2022)

Factors	Example Item
Story	A series of events happened in the game that told a story.
Improving skills	My ability to play this game well improved the more that I played the game.
Cooperation	My teammates and I worked well together as a team to achieve our objectives. The teammates could be other players in the real world or characters in the game.
Altered perception of time	I was not aware of how much time had passed until after the game was over.
Optimal pacing	The speed of the game was just right for me – not too slow and not too fast.
Theme	I liked the theme or topic of the game. The theme or topic of the game is the main subject matter the game focused on, such as fishing, ancient Egypt, or boxing.
Relaxation	I had relaxing experiences.
Ability to retry	I was able to retry what I was doing if I did not successfully complete the task.
Humor	I burst out laughing.
Organizing	I made things well-organized.
Feedback	The game gave me feedback that made it really clear to me how well I was doing at playing the game.
Usability of controls	The controls of the game were easy to learn.
Focusing of full attention	I had my attention focused entirely on what I was doing.
Autonomy	I was deciding for myself how I would take action.
Exercise	I was moving my body enough to make my heart beat faster.
Competition	I felt that I had more skill at playing the game than other people. The other people could be other players in the real world or characters in the game.
Honor	What I was doing in the game was in accordance with my personal moral standards.
Creating	I combined different parts to make new things.
Suspense	I felt suspense because I did not know what was going to happen next.
Vitality	I felt energized.
Merging of action and awareness	My actions in the game just seemed to happen automatically.
Learning	I learned a lot while playing this game.
Schadenfreude-vengeance	I got even with other players or characters in the game who offended me.
Leading-directing	I was the leader with the authority to direct the actions of others. The others could be other players in the real world or characters in the game.
Sensory pleasure and beauty	What I saw and heard in this game was beautiful.
Presence	I felt as if I was my character.
Clear task purpose	It was clear to me why what I was doing mattered in the game.
Strategizing	I thought up a plan to take action in the game.
Social responsibility	It felt like what I was doing was benefiting society. The society could be in the game world or in the real world.

(Continued)

TABLE 11.1 (Continued)
Factors of the Sources of Enjoyment Questionnaire (Schaffer, 2022)

Factors	Example Item
Optimal challenge	Playing this game was challenging for me, but I was able to face that challenge.
Loss of self-consciousness	I was not thinking about how I was presenting myself.
Achievement	I felt proud of myself when I succeeded in the game.
Clear goals and navigation	It was clear to me what to do next through each step of the game.
Sense of control	I felt as though I had everything under control.
Goal attainability	I felt that my current goal in the game was achievable.
Optimal variety	The game had a variety of new content as I went through it, but not so much variety that it was overwhelming.
Savoring	I took the time to savor what I was experiencing.
Task significance	What I was doing in this game was meaningful.

atomic actions. Atomic actions are the smallest action steps that cannot be broken down further into additional feedback loops within feedback loops.

The feedback loop of flow model aligns with qualitative descriptions of the flow process. Nakamura and Csikszentmihalyi described flow as "engaging just-manageable challenges by tackling a series of goals, continuously processing feedback about progress, and adjusting action based on this feedback" (Nakamura & Csikszentmihalyi, 2014, p. 90). The feedback loop of flow model captures this process by showing the steps of perception, cognition, and action that come together to create the flow experience, and how the flow conditions provide the inputs needed for that cognitive process.

EMPIRICAL INVESTIGATIONS OF DIGITAL GAME ENJOYMENT

LITERATURE REVIEW AS FOUNDATION

This empirical investigation into the sources of digital game enjoyment began with a comprehensive literature review that drew from multiple academic disciplines. This interdisciplinary approach incorporated theories and findings from psychology, game design, anthropology, philosophy, information systems, and HCI. By examining this diverse body of knowledge, the research aimed to identify a comprehensive set of factors that have been theorized or empirically demonstrated to contribute to positive gameplay experiences.

The main results of this literature review are presented in section "Theoretical Foundations of Digital Game Enjoyment", outlining a wide variety of empirical studies that informed this investigation. See Schaffer and Fang for an extended review of previous research on digital game enjoyment (Schaffer & Fang, 2019). This literature review identified many potential sources of enjoyment, suggesting that digital game enjoyment may emerge from a wide variety of psychological, social, and system design factors. These factors laid the theoretical groundwork for the empirical investigation of positive player experiences described in the following sections.

Card Sorting Study with Game Players

To build on the insights gained from the literature review and to ground the investigation in the lived experiences of game players, an iterative card sorting study was conducted by Schaffer and Fang (2018). 60 participants were recruited who reported playing digital games at least once per week, ensuring that the perspectives of active gamers were central to the research.

The study began with a set of 167 potential sources of enjoyment, derived primarily from the extensive literature reviewed in the preceding stage. The core task for participants was to sort these individual sources of enjoyment, each presented on a separate card, into categories. This categorization process was iterative, and the cards and categories were revised after every 10 participants. The first 40 participants had the flexibility to place cards into multiple categories, create entirely new categories, and designate sources as not being potential sources of digital game enjoyment. This open-ended approach allowed for the emergence of player-defined conceptualizations of enjoyment.

For the final 20 participants, the level of agreement between participants regarding the categorization of cards was assessed using Randolph's multi-rater kappa (Randolph, 2005), which yielded a score of 0.93, indicating a very high level of consensus. This card sorting procedure mirrored methods employed in the development of other psychological measures, such as those used by Moore and Benbasat (1991), and in previous research on enjoyment and flow in computer games (Fang et al., 2010, 2013).

The outcome of this iterative sorting process was a refined set of 34 distinct categories representing the sources of enjoyment in digital games, grounded both in existing theoretical frameworks and in the direct experiences of game players. This process identified novel sources of enjoyment not previously highlighted in game enjoyment research, such as savoring, humor, and relaxation.

Development of Measurement Instruments

Following the identification of the 34 sources of enjoyment through the card sorting study, the subsequent phase of the research focused on developing psychometrically sound instruments to measure these constructs. Building directly on the 34 categories, Schaffer developed the Enjoyment Questionnaire (EQ) to provide a measure of enjoyment and the Sources of Enjoyment Questionnaire (SoEQ) to assess how much players experience each of the identified sources of enjoyment (Schaffer, 2022).

The initial step in this development process involved item generation, where a pool of 6–22 items was devised for each of the 34 sources of enjoyment and for overall enjoyment, adhering to established guidelines for scale development (DeVellis & Thorpe, 2021), which recommend generating a larger initial item pool. Preliminary validation of these newly developed measures was conducted through a survey of

564 participants who had played a digital game within the last six months, asking them to reflect on their most recent experience playing a digital game.

Factor analysis, a statistical technique used to identify underlying dimensions within a set of variables, was then conducted to refine the scales. Through an iterative process of examining factor loadings and cross-loadings, items that did not align strongly with a single intended factor or that exhibited substantial relationships with multiple factors were systematically removed. This process sometimes resulted in the merging or splitting of intended factors, leading to a final set of 38 factors in the SoEQ (see Table 11.1).

The resultant factor structure demonstrated strong psychometric properties. Factor loadings exceeded |0.4| for convergent validity, and cross-loadings were at least 0.2 lower than primary loadings for discriminant validity (Hair et al., 2019). All scales achieved Cronbach's alpha levels above 0.7, with all but one scale (Loss of Self-Consciousness) exceeding 0.8, indicating high internal consistency reliability (DeVellis & Thorpe, 2021). These results supported the construct validity and reliability of both the EQ and the SoEQ.

To accommodate diverse research needs, Schaffer (2022) developed short and long versions of both questionnaires. The short versions were created by reducing the number of items per construct to a maximum of four, aiming for a more parsimonious instrument suitable for applied research contexts where minimizing participant burden is important. Conversely, the longer versions are recommended for basic research endeavors or when researchers seek to confirm the factor structure of the measures in different populations or contexts.

Researchers can either select specific subscales of the SoEQ relevant to their research questions or administer the complete measure for a more comprehensive assessment of the various sources of enjoyment. Participants rate their agreement with each statement on a 7-point Likert scale after playing a game or other interactive experience, and the mean scores provide measures of overall enjoyment (with the EQ) and the degree to which each source of enjoyment was experienced (with the SoEQ).

EVALUATION METHODS FOR PLAYER EXPERIENCE

RESEARCH OBJECTIVES AND APPROACHES

The evaluation of player experience in digital games is guided by clearly defined research objectives, which generally fall into two primary categories: applied research and basic research.

Applied research is concerned with the practical goal of improving specific games or interactive systems currently under development. The focus is on identifying issues that hinder enjoyment or discovering ways to make the experience more positive for a particular project.

In contrast, basic research aims to generate generalizable knowledge about the fundamental factors that contribute to positive player experiences. The insights gained from basic research can then inform applied research and the design of future enjoyable interactive systems.

It is important to select research methods that align well with the specific research objectives. The questions researchers seek to answer will determine whether qualitative, quantitative, or mixed-methods approaches are most suitable for the investigation.

APPLIED RESEARCH METHODS

Applied player experience research often involves extending traditional usability testing methodologies to specifically address aspects of enjoyment rather than solely focusing on ease of use. While usability testing aims to identify problems users encounter while interacting with a system, applied player experience evaluation expands this to identify elements that detract from or contribute to a positive experience.

Common techniques adapted from usability testing include think-aloud protocols, where participants verbalize their thoughts and feelings as they play, participant observation of player behavior during gameplay, and post-experience interviews, which allow researchers to gather qualitative data about the player's overall experience and specific moments of enjoyment or frustration.

The Microsoft Games User Research group has developed several noteworthy methods tailored for evaluating and improving games during development. The Rapid Iterative Testing and Evaluation (RITE) method emphasizes making immediate changes to the game as issues are identified and continuing testing with revised versions, often involving multiple short usability tests with small numbers of participants (Medlock et al., 2002). This iterative approach allows for quick identification and resolution of problems.

The playtest method combines usability testing with surveys administered to a larger group of participants, typically 25–35 participants (Davis et al., 2005). This larger sample size allows for null-hypothesis significance testing to compare different versions of a game or different games of the same genre to determine if design changes have a statistically significant impact on player enjoyment. It is worth noting that the term playtest has become a more general term for user testing games.

Finally, the Tracking Real-time User Experience (TRUE) system combines automatic player behavior tracking, video recording, and short in-game surveys that are synchronized by timestamps (Kim et al., 2008; Schuh et al., 2008). This allows researchers to identify exact moments in the gameplay videos that match specific behavioral data and survey responses, providing detailed insights into why players experience particular problems or positive experiences. However, setting up systems like TRUE requires a significant initial investment of time and resources.

BASIC RESEARCH METHODS

We present an input-process-output model of digital game enjoyment as a way to structure player experience research (see Figure 11.2). In this framework, inputs are the sources of enjoyment that games provide, the process includes the cognitive and behavioral steps players go through while experiencing enjoyment during gameplay, and outputs are the benefits and lasting effects that result from enjoyable gaming

Input		Process		Output
Sources of Enjoyment	→	Steps and Experience of Enjoyment	→	Outcomes and Benefits of Enjoyement

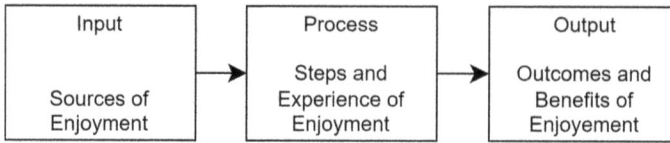

FIGURE 11.2 Input-process-output model of digital game enjoyment.

experiences. This approach allows researchers to examine what makes games enjoyable, how the steps and experience of enjoyment unfold during gameplay, and what benefits players gain from digital game enjoyment.

Basic player experience research uses various methods to identify new sources of enjoyment and understand how these sources impact digital game enjoyment. Qualitative approaches are important in the initial stages of identifying potential sources of enjoyment. Methods such as interviews, surveys with open-ended questions, and card sorting studies help researchers understand players' emotions, needs, and desires related to gameplay, as well as explore the cognitive and emotional processes players experience during enjoyable moments of gameplay.

For example, Schaffer and Fang (2018) conducted a card sorting study where players categorized 167 potential sources of enjoyment drawn from a literature review, identifying 34 distinct categories. The first 40 participants were able to add sources of enjoyment they felt were missing and sort cards into a "Not a Potential Source of Enjoyment" category. The cards and categories were revised after every 10 participants. By the final two rounds, participants achieved very high inter-rater reliabilities of 0.9381 and 0.9367, as calculated with Randolph's free-marginal multi-rater kappa (Randolph, 2005), indicating strong agreement in how they categorized the sources. This iterative approach revealed novel sources of enjoyment such as savoring, humor, and relaxation, demonstrating the value of qualitative methods in uncovering the diverse sources that contribute to player enjoyment.

To test relationships between identified sources of enjoyment and player outcomes, controlled experiments are particularly well-suited. These studies create multiple versions of the same game with targeted design differences, randomly assign participants to different versions, and use post-game surveys to measure impacts on enjoyment. The Enjoyment Questionnaire (EQ) and Sources of Enjoyment Questionnaire (SoEQ) (Schaffer, 2022) can be used to measure 38 sources of enjoyment and how much players experienced enjoyment itself (see Table 11.1).

Designing effective controlled experiments involves balancing how much the research game reflects commercially available digital games that people play with the need for consistency across participants to isolate the effects of the manipulated variables. An example of such research is the controlled experiment by Ambati, Brandt, and Schaffer, which investigated the impact of a navigational pointer, a quest log, and humorous Non-Player Character (NPC) dialog on player enjoyment and flow in a custom-built action role-playing game (Ambati et al., 2022). The study found that a navigational pointer significantly increased enjoyment and flow, while the quest log and humorous dialog did not have a significant impact. This research demonstrates how specific game features, such as navigational aids, can influence how much players experience flow and enjoyment. This highlights the importance of testing specific design choices to understand their effects.

Effective player experience research begins with clear, specific research questions that define what sources of enjoyment will be studied. When conducting qualitative research, it is best to use systematic data analysis methods to examine interview transcripts and identify patterns in the data. Two recommended approaches are grounded theory (Corbin & Strauss, 2014) and thematic analysis (Clarke et al., 2015). These methods help researchers code their data and identify recurring themes across different sources of player enjoyment.

APPLICATIONS OF PLAYER EXPERIENCE RESEARCH

GAME DESIGN APPLICATIONS

Player experience research provides game developers with concrete insights for creating games that players will want to play and will keep them coming back for more. Since the main goal of most game developers is to create positive, enjoyable experiences that lead to user engagement and retention, understanding the sources of enjoyment in digital games is important for making informed design decisions. By recognizing how different design elements contribute to positive player experiences, developers can make more evidence-based decisions throughout development. This empirical approach moves beyond intuition and industry conventions, using research findings about player psychology to guide creative choices and effectively design for player enjoyment, engagement, and retention.

Research has identified 38 sources of enjoyment in digital games, the factors in the Sources of Enjoyment Questionnaire (SoEQ; Schaffer, 2022). This framework provides developers with specific targets to design for during ideation and development. Game designers and developers can use this framework by strategically choosing sources of enjoyment to design for, mixing and combining them the way a painter mixes and combines colors with a palette to achieve the desired experience. Designing for these sources of enjoyment creates experiences that more effectively satisfy player needs and desires to maximize enjoyment and sustain engagement.

Designing for flow in games is a core part of applied player experience research. Flow theory suggests that optimal challenges, clear proximal goals, and immediate progress feedback are the conditions that lead to a state of flow, characterized by deep engagement and enjoyment. By designing for these flow conditions, game designers can structure gameplay to facilitate players getting into a flow state. For example, implementing dynamic difficulty adjustment by having the difficulty of the game increase and decrease according to player performance can help maintain an optimal level of challenge. Similarly, providing clear objectives for the immediate next step of the activity and offering continuous feedback on how well players are playing the game can help players get into and stay in a flow state. When these flow conditions come together, they create a feedback loop of flow where the player continually cycles through goal setting, testing progress, and taking action to overcome optimally challenging tasks (Schaffer & Fang, 2022).

The EQ and the SoEQ represent practical tools for assessing and improving player experiences in game development (Schaffer, 2022). These questionnaires allow researchers and practitioners to measure overall enjoyment and the extent to which players experience 38 specific sources of enjoyment (shown in Table 11.1). By administering these questionnaires to players during playtesting or after they have played a game, developers can gather quantitative data on how well their design choices contribute to enjoyment. This data can then inform iterative design processes, allowing developers to refine gameplay mechanics, narrative elements, or other game design elements or features to improve the player experience based on empirical evidence. Applied researchers might prefer the shorter versions of the EQ and SoEQ to minimize participant fatigue during testing.

SERIOUS GAMES AND GAMIFICATION

The methods and best practices from player experience research apply directly to serious games and gamification. Serious games are "full-fledged games for non-entertainment purposes," while gamification means "the use of game design elements in non-game contexts" (Deterding et al., 2011). Both approaches try to harness the engaging qualities of games to reach specific goals beyond entertainment.

When serious games include elements that create positive player experiences, they become more effective at reaching their intended goals, whether those goals involve education, research, health-related behavior change, or persuasion. Similarly, gamification can boost user engagement and motivation by making non-game systems more game-like and enjoyable. Understanding what makes games enjoyable helps designers choose the game elements best suited for their specific purposes.

Several real-world examples of successful serious games demonstrate how these methods work in practice. *Foldit* focuses on protein folding research and citizen science by turning complex biological prediction problems into engaging gameplay (Cooper et al., 2010; Koepnick et al., 2019). The game uses crowdsourcing to harness the collective problem-solving abilities of many players for biochemistry research. By transforming a scientific challenge into an interactive puzzle, it draws on players' problem-solving skills while giving them a sense of achievement and contribution.

EndeavorRx represents a breakthrough in healthcare gaming as the first prescription video game approved by the United States Food and Drug Administration (US FDA) for treating Attention-Deficit/Hyperactivity Disorder (ADHD) in children ages 8–17 (Kollins et al., 2020, 2021). It demonstrates how game mechanics can serve therapeutic purposes by training players to ignore distractions while multitasking, with this game using dynamic difficulty adjustment and progress rewards to keep players engaged. The design taps into enjoyment sources such as focusing of full attention, optimal challenge, feedback, and achievement.

Survive the Century is a serious game designed to educate the public about climate change (Beckbessinger et al., n.d.). Players make political, environmental, and social choices about adapting to climate change from 2021 to 2100. The game makes use of sources of enjoyment, including story, social responsibility, and suspense, as players navigate uncertain futures.

From crowdsourced scientific research to US FDA-approved medical treatments to climate change education, these cases illustrate how understanding player experience can transform serious real-world problems into engaging interactive experiences that people want to play. The main lesson is that successful serious games strategically select sources of enjoyment based on what they're trying to accomplish, whether that's getting lots of people to work together to solve puzzles for research, helping people with ADHD improve their focus and attention, or getting people to care more about the environment and take climate action.

ETHICAL CONSIDERATIONS IN DESIGNING FOR ENJOYMENT

As researchers and practitioners become effective at designing for enjoyment, we need to consider the ethical implications of the interactive systems we create. Games and other interactive systems have the power to shape user experiences and motivate action. With that power comes the responsibility to improve users' lives and make the world a better place. Interactive systems we design need to provide experiences and motivate actions that align with user values, needs, and desires. The focus needs to be on creating positive, enjoyable experiences that users genuinely feel are worthwhile and want to engage with.

The experiences we create and actions our systems motivate must follow ethical principles of beneficence, respect for autonomy, non-maleficence, and justice. Beneficence means actively promoting users' well-being and joy. Respect for autonomy involves allowing people to make voluntary, informed decisions by providing clear, accurate information and meaningful control. Non-maleficence involves anticipating and preventing harm or unintended side effects to prioritize users' safety and trust. Justice calls for fairness in access and outcomes, ensuring every user benefits equally from our work regardless of background, ability, or circumstance. These four principles are the foundation of biomedical ethics and provide a solid framework for ethical design when interactive systems impact human well-being.

Designing for enjoyment means creating positive experiences people want to have, not designing systems that use manipulative techniques to form addictive habits or control behavior. This distinction is especially important when applying player experience research and methods to serious games or gamification. We need to avoid slot machine-style external rewards that rely on behaviorist variable-ratio reward schedules to create addictive habit formation (Schüll, 2012). The goal is to design for intrinsic motivation that comes from enjoying the activity itself, rather than relying on external rewards that can undermine user autonomy and well-being.

Enjoyable systems are powerful because they can motivate people to take action. We need to use this motivational power to improve people's lives and advance important goals like health, education, and improving society. Ethical design means asking whether the systems we build truly benefit the people who use them. We must consider ethics throughout the design process to ensure the experiences we create serve users' best interests and align with these positive goals.

CONCLUSION AND FUTURE DIRECTIONS

For those who want to design and develop enjoyable interactive systems, this chapter offers practical methods and tools. Research from psychology, game design, anthropology, philosophy, information systems, and HCI was reviewed to understand what makes games enjoyable. Measurement tools were then developed and tested so that researchers and designers can use them to create better interactive experiences.

The feedback loop of flow model shows how clear proximal goals, immediate progress feedback, and optimal challenges create the conditions that get players into a flow state. Building on cognitive psychology's TOTE units as the fundamental building blocks of cognition, the feedback loop of flow model extends this process model by adding a goal-setting step before the test-operate-test-exit cycle. This creates a recursive process where players set goals, test progress, take action, and test again until goals are achieved, operating at multiple levels of task abstraction simultaneously.

The literature review revealed that people enjoy games for many different reasons. A card sorting study with 60 game players was conducted to build on these theoretical insights and ground the investigation in the lived experiences of active gamers. Through an iterative process that achieved high inter-rater agreement, this study identified distinct categories of enjoyment, including novel sources not previously highlighted in game enjoyment research, such as savoring, humor, and relaxation.

Two measurement tools were developed based on the sources of enjoyment identified in the card sorting study: the Enjoyment Questionnaire (EQ) and Sources of Enjoyment Questionnaire (SoEQ). During the development process, factor analysis refined the original 34 categories into 38 distinct sources of enjoyment that give researchers and designers specific targets to aim for (see Table 11.1 for the full list of 38 sources of enjoyment). These questionnaires performed well in testing, with strong factor loadings and reliability scores above 0.7 for all scales. An input-process-output model was also presented that helps researchers investigate how design choices lead to positive experiences. These tools work well for both applied researchers improving specific games and basic researchers exploring digital game enjoyment and its sources.

The applications go far beyond entertainment. Successful serious games demonstrate how understanding player experience can transform real-world challenges into engaging interactive experiences across domains including scientific research, healthcare, and education. These examples show how strategic selection of enjoyment sources based on specific goals can make serious games more effective at achieving their intended outcomes.

The ethics of designing for enjoyment were also addressed. The key is focusing on designing for intrinsic motivation rather than using extrinsic rewards to control users. The ethical framework presented, based on principles of beneficence, autonomy, non-maleficence, and justice, helps designers create systems that genuinely benefit users.

This work opens up many opportunities for future research. Controlled experiments are needed to provide evidence about which design elements actually increase enjoyment. Long-term studies are needed to see how positive gameplay experiences affect outcomes such as learning, behavior change, meaning, and happiness over time. Research across different cultures is needed to understand how sources of enjoyment vary between populations.

Most importantly, researchers need to bridge theory and practice by translating the 38 sources of enjoyment into concrete design implementations. This means figuring out specific ways to design for these sources of enjoyment in games, such as through mechanics, narratives, user-interface elements, and design choices. For example, the "clear goals and navigation" source of enjoyment was implemented effectively through a directional arrow on screen that guides players toward their next objective (Ambati et al., 2022). Similar concrete implementations need to be identified for each source of enjoyment and tested in controlled experiments.

The field of player experience research addresses fundamental questions about human motivation and well-being in interactive systems. With validated measurement tools, established theoretical frameworks, and growing evidence of real-world applications, the foundation exists for creating meaningful interactive experiences. The methods are proven, the opportunities are clear, and the potential impact is significant. The next generation of researchers and designers can build on this work to develop systems that not only engage users but also meaningfully improve their lives and help make the world a better, more enjoyable place.

REFERENCES

Abeele, Vero Vanden, Spiel, Katta, Nacke, Lennart, Johnson, Daniel, & Gerling, Kathrin. (2020). Development and validation of the player experience inventory: A scale to measure player experiences at the level of functional and psychosocial consequences. *International Journal of Human-Computer Studies*, *135*, 102370. https://doi.org/10.1016/j.ijhcs.2019.102370

Ambati, Uday Sai Reddy, Brandt, Gregory, & Schaffer, Owen. (2022). Guidance is Good: Controlled Experiment Shows the Impact of Navigational Guidance on Digital Game Enjoyment and Flow. In Xiaowen Fang (Ed.), *HCI in Games* (pp. 101–118). Springer International Publishing. https://doi.org/10.1007/978-3-031-05637-6_7

American Psychiatric Association. (2013). *Diagnostic and statistical manual of mental disorders (DSM-5®)*. American Psychiatric Pub.

Bartle, Richard. (1996). Hearts, clubs, diamonds, spades: Players who suit MUDs. *Journal of MUD Research*, *1*(1), 19.

Beckbessinger, Sam, Nicholson, Simon, & Trisos, Christopher. (n.d.). *Survive the Century*. Retrieved June 17, 2025, from https://survivethecentury.net/

Boberg, Marion, Karapanos, Evangelos, Holopainen, Jussi, & Lucero, Andrés. (2015). PLEXQ: Towards a Playful Experiences Questionnaire. In *Proceedings of the 2015 Annual Symposium on Computer-Human Interaction in Play*, 381–391. https://doi.org/10.1145/2793107.2793124

Brown, Stuart, & Vaughan, Christopher. (2010). *Play: How it Shapes the Brain, Opens the Imagination, and Invigorates the Soul* (Reprint edition). Avery.

Caillois, Roger. (1961). *Man, Play, and Games*. University of Illinois Press.

Clarke, Victoria, Braun, Virginia, & Hayfield, Nikki. (2015). Thematic analysis. *Qualitative Psychology: A Practical Guide to Research Methods*, *222*(2015), 248.

Condon, P., Wilson-Mendenhall, C. D., & Barrett, L. F. (2014). What is a Positive Emotion?: The Psychological Construction of Pleasant Fear and Unpleasant Happiness. In M. M. Tugade, M. N. Shiota, & L. D. Kirby (Eds.), *Handbook of Positive Emotions* (pp. 60–81). The Guilford Press

Cooper, Seth, Khatib, Firas, Treuille, Adrien, Barbero, Janos, Lee, Jeehyung, Beenen, Michael, Leaver-Fay, Andrew, Baker, David, Popović, Zoran, & Players, Foldit. (2010). Predicting protein structures with a multiplayer online game. *Nature*, *466*(7307), Article 7307. https://doi.org/10.1038/nature09304

Corbin, Juliet, & Strauss, Anselm. (2014). *Basics of Qualitative Research: Techniques and Procedures for Developing Grounded Theory* (4 edition). SAGE Publications, Inc.

Costello, Brigid, & Edmonds, Ernest. (2007). A Study in Play, Pleasure and Interaction Design. In *Proceedings of the 2007 Conference on Designing Pleasurable Products and Interfaces*, 76–91. http://dl.acm.org/citation.cfm?id=1314168

Csıkszentmihályi, Mihaly. (1975). *Beyond Boredom and Anxiety*. Jossey-Bass Inc Pub.

Csikszentmihalyi, Mihaly. (1993). *The Evolving Self: A Psychology for the Third Millennium*. HarperCollins.

Csikszentmihalyi, Mihaly. (1998). *Finding Flow: The Psychology of Engagement with Everyday Life* (1st edition). Basic Books.

Csikszentmihalyi, Mihaly. (2008). *Flow: The Psychology of Optimal Experience* (1 edition). Harper Perennial Modern Classics.

Csikszentmihalyi, Mihaly, Abuhamdeh, Sami, & Nakamura, Jeanne. (2014). Flow. In Mihaly Csikszentmihalyi (Ed.), *Flow and the Foundations of Positive Psychology* (pp. 227–238). Springer Netherlands. https://doi.org/10.1007/978-94-017-9088-8_15

Csikszentmihalyi, Mihaly, & Csikszentmihalyi, Isabella Selega (Eds.). (1988). *Optimal Experience: Psychological Studies of Flow in Consciousness*. Cambridge University Press.

Csikszentmihalyi, Mihaly, & Nakamura, Jeanne. (2010). Effortless Attention in Everyday Life: A Systematic Phenomenology. In B. Bruya (Ed.), *Effortless Attention: A New Perspective in the Cognitive Science of Attention and Action* (pp. 179–189). Boston Review.

Davis, John P., Steury, Keith, & Pagulayan, Randy. (2005). A survey method for assessing perceptions of a game: The consumer playtest in game design. *Game Studies*, *5*(1), 1–13.

Deci, Edward, & Ryan, Richard M. (1985). *Intrinsic Motivation and Self-Determination in Human Behavior*. Plenum Press.

Deterding, Sebastian, Dixon, Dan, Khaled, Rilla, & Nacke, Lennart. (2011). From Game Design Elements to Gamefulness: Defining Gamification. In *Proceedings of the 15th International Academic MindTrek Conference: Envisioning Future Media Environments*, 9–15. http://dl.acm.org/citation.cfm?id=2181040

DeVellis, Robert F., & Thorpe, Carolyn T. (2021). *Scale Development: Theory and Applications*. Sage Publications.

Fang, Xiaowen, Chan, Susy, Brzezinski, Jacek, & Nair, Chitra. (2010). Development of an instrument to measure enjoyment of computer game play. *International Journal of Human-Computer Interaction*, *26*(9), 868–886. https://doi.org/10.1080/10447318.201 0.496337

Fang, Xiaowen, Zhang, Jingli, & Chan, Susy S. (2013). Development of an instrument for studying flow in computer game play. *International Journal of Human-Computer Interaction*, *29*(7), 456–470. https://doi.org/10.1080/10447318.2012.715991

Fredrickson, Barbara. (2009). *Positivity: Top-Notch Research Reveals the Upward Spiral That Will Change Your Life* (1 edition). Harmony.

Fullerton, Tracy. (2014). *Game Design Workshop: A Playcentric Approach to Creating Innovative Games, Third Edition* (3 edition). A K Peters/CRC Press.

Hair, Joseph F., Black, William C., Babin, Barry J., & Anderson, Rolph E. (2019). *Multivariate Data Analysis* (8th edition). Cengage Learning.

Havercamp, Susan M. (1998). *The Reiss Profile of motivation sensitivity: Reliability, validity, and social desirability* [PhD Thesis]. Ohio State University.

Hektner, Joel M., Schmidt, Jennifer A., & Csikszentmihalyi, Mihaly. (2007). *Experience Sampling Method: Measuring the Quality of Everyday Life*. Sage.

Jackson, Susan A., & Eklund, Robert Charles. (2004). *The Flow Scales Manual*. Fitness Information Technology.

Jackson, Susan A., & Marsh, Herbert W. (1996). Development and validation of a scale to measure optimal experience: The Flow State Scale. *Journal of Sport and Exercise Psychology*, *18*(1), 17–35.

Kim, Jun H., Gunn, Daniel V., Schuh, Eric, Phillips, Bruce, Pagulayan, Randy J., & Wixon, Dennis. (2008). Tracking Real-Time User Experience (TRUE): A Comprehensive Instrumentation Solution for Complex Systems. In *Proceedings of the SIGCHI Conference on Human Factors in Computing Systems*, 443–452. https://doi.org/10.1145/1357054.1357126

Koepnick, Brian, Flatten, Jeff, Husain, Tamir, Ford, Alex, Silva, Daniel-Adriano, Bick, Matthew J., Bauer, Aaron, Liu, Gaohua, Ishida, Yojiro, Boykov, Alexander, Estep, Roger D., Kleinfelter, Susan, Nørgård-Solano, Toke, Wei, Linda, Players, Foldit, Montelione, Gaetano T., DiMaio, Frank, Popović, Zoran, Khatib, Firas, … Baker, David. (2019). De novo protein design by citizen scientists. *Nature*, *570*(7761), Article 7761. https://doi.org/10.1038/s41586-019-1274-4

Kollins, Scott H., Childress, Ann, Heusser, Andrew C., & Lutz, Jacqueline. (2021). Effectiveness of a digital therapeutic as adjunct to treatment with medication in pediatric ADHD. *NPJ Digital Medicine*, *4*(1), 58.

Kollins, Scott H., DeLoss, Denton J., Cañadas, Elena, Lutz, Jacqueline, Findling, Robert L., Keefe, Richard SE, Epstein, Jeffery N., Cutler, Andrew J., & Faraone, Stephen V. (2020). A novel digital intervention for actively reducing severity of paediatric ADHD (STARS-ADHD): A randomised controlled trial. *The Lancet Digital Health*, *2*(4), e168–e178.

Korhonen, Hannu, Montola, Markus, & Arrasvuori, Juha. (2009). Understanding Playful User Experience Through Digital Games. In *International Conference on Designing Pleasurable Products and Interfaces, 2009*. http://citeseerx.ist.psu.edu/viewdoc/download?doi=10.1.1.586.7146&rep=rep1&type=pdf

Lazzaro, N. (2004). *Why We Play Games: Four Keys to More Emotion Without Story*. http://gamemodworkshop.com/readings/xeodesign_whyweplaygames.pdf

Lazzaro, Nicole. (2009). Why We Play: Affect and the Fun of Games. In A. Sears & J. A. Jacko (Eds.), *Human-Computer Interaction: Designing for Diverse Users and Domains* (pp. 155–176). CRC Press.

Lucero, Andrés, & Arrasvuori, Juha. (2010). PLEX Cards: A Source of Inspiration When Designing for Playfulness. In *Proceedings of the 3rd International Conference on Fun and Games*, 28–37. https://doi.org/10.1145/1823818.1823821

Lucero, Andrés, & Arrasvuori, Juha. (2013). The PLEX Cards and its techniques as sources of inspiration when designing for playfulness. *International Journal of Arts and Technology*, *6*(1), 22–43.

Medlock, Michael C., Wixon, Dennis, Terrano, Mark, Romero, Ramon, & Fulton, Bill. (2002). Using the RITE method to improve products: A definition and a case study. *Usability Professionals Association*, *51*, 1963813932.

Miller, George A., Galanter, Eugene, & Pribram, Karl H. (1960). *Plans and the Structure of Behavior*. Martino Fine Books.

Moore, Gary C., & Benbasat, Izak. (1991). Development of an instrument to measure the perceptions of adopting an information technology innovation. *Information Systems Research*, *2*(3), 192–222.

Nakamura, J., & Csikszentmihalyi, M. (2014). The Concept of Flow. In *Flow and the Foundations of Positive Psychology* Springer, Dordrecht. https://doi.org/10.1007/978-94-017-9088-8_16

Park, Nansook, Peterson, Christopher, & Seligman, Martin E. P. (2004). Strengths of character and well-being. *Journal of Social and Clinical Psychology*, *23*(5), 603–619.

Peterson, Christopher, Park, Nansook, & Seligman, Martin EP. (2005). Orientations to happiness and life satisfaction: The full life versus the empty life. *Journal of Happiness Studies*, *6*(1), 25–41.

Peterson, Christopher, & Seligman, Martin EP. (2004). *Character Strengths and Virtues: A Handbook and Classification* (Vol. 1). Oxford University Press.

Quick, John M., Atkinson, Robert K., & Lin, Lijia. (2012). Empirical taxonomies of gameplay enjoyment: personality and video game preference. *International Journal of Game-Based Learning*, *2*(3), 11–31.

Randolph, Justus J. (2005). *Free-marginal multirater kappa: An alternative to Fleiss' fixed-marginal multirater kappa*. Joensuu University Learning and Instruction Symposium 2005, Joensuu, Finland. https://eric.ed.gov/?id=ED490661

Reiss, Steven. (2002). *Who Am I? The 16 Basic Desires That Motivate Our Actions and Define Our Personalities* (35050th edition). Berkley.

Reiss, Steven. (2004). Multifaceted nature of intrinsic motivation: the theory of 16 basic desires. *Review of General Psychology*, *8*(3), 179–193. https://doi.org/10.1037/1089-2680.8.3.179

Reiss, Steven, & Havercamp, Susan M. (1998). Toward a comprehensive assessment of fundamental motivation: Factor structure of the Reiss Profiles. *Psychological Assessment*, *10*(2), 97–106. https://doi.org/10.1037/1040-3590.10.2.97

Ryan, R. M., & Deci, E. L. (2000). Self-determination theory and the facilitation of intrinsic motivation, social development, and well-being. *American Psychologist*, *55*(1), 68–78. https://doi.org/10.1037//0003-066X.55.1.68

Ryan, R. M., Rigby, S. C., & Przybylski, A. (2006). The motivational pull of video games: a self-determination theory approach. *Motivation and Emotion*, *30*(4), 344–360. https://doi.org/10.1007/s11031-006-9051-8

Schaffer, Owen. (2019). *A desire fulfillment theory of digital game enjoyment* [Ph.D. Dissertation, DePaul University]. College of Computing and Digital Media Dissertations. 18. https://via.library.depaul.edu/cdm_etd/18

Schaffer, Owen. (2022). Development and Preliminary Validation of the Enjoyment Questionnaire and the Sources of Enjoyment Questionnaire. In *Extended Abstracts of the 2022 CHI Conference on Human Factors in Computing Systems*, 1–7. https://doi.org/10.1145/3491101.3519819

Schaffer, Owen, & Fang, Xiaowen. (2018). *What Makes Games Fun? Card Sort Reveals 34 Sources of Computer Game Enjoyment*. Americas Conference on Information Systems (AMCIS) 2018, New Orleans. http://aisel.aisnet.org/amcis2018/HCI/Presentations/2/

Schaffer, Owen, & Fang, Xiaowen. (2019). Digital Game Enjoyment: A Literature Review. In *HCI in Games*, 191–214. https://doi.org/10.1007/978-3-030-22602-2_16

Schaffer, Owen, & Fang, Xiaowen. (2022). The feedback loop of flow: controlled experiment shows task-relevant feedback increases flow. *AIS Transactions on Human-Computer Interaction*, *14*(3), 356–389. https://doi.org/10.17705/1thci.00172

Schuh, Eric, Gunn, Daniel V., Phillips, Bruce, Pagulayan, Randy J., Kim, Jun H., & Wixon, Dennis. (2008). Chapter 15—TRUE Instrumentation: Tracking Real-Time User Experience in Games. In *Game Usability* (pp. 237–265). Morgan Kaufmann. https://doi.org/10.1016/B978-0-12-374447-0.00015-9

Schüll, Natasha Dow. (2012). *Addiction by Design: Machine Gambling in Las Vegas*. Princeton University Press. https://doi.org/10.1515/9781400834655

Seligman, Martin E. P., & Csikszentmihalyi, Mihaly. (2000). Positive psychology: An introduction. *American Psychologist*, *55*(1), 5–14. https://doi.org/10.1037/0003-066X.55.1.5

Shiota, Michelle N. (2014). The Evolutionary Perspective in Positive Emotion Research. In M. M. Tugade, M. N. Shiota, & L. D. Kirby (Eds.), *Handbook of Positive Emotions* (pp. 44–59). The Guilford Press.

Shiota, Michelle N., Neufeld, Samantha L., Danvers, Alexander F., Osborne, Elizabeth A., Sng, Oliver, & Yee, Claire I. (2014). Positive emotion differentiation: a functional approach: positive emotion differentiation. *Social and Personality Psychology Compass*, *8*(3), 104–117. https://doi.org/10.1111/spc3.12092

Sutton-Smith, Brian. (2009). *The Ambiguity of Play*. Harvard University Press.

Yee, Nick. (2006). Motivations for play in online games. *CyberPsychology & Behavior*, *9*(6), 772–775.

Yee, Nick. (2016). The Gamer Motivation Profile: What We Learned From 250,000 Gamers. In *Proceedings of the 2016 Annual Symposium on Computer-Human Interaction in Play*, 2–2. https://doi.org/10.1145/2967934.2967937

Yee, Nick, Ducheneaut, Nicolas, & Nelson, Les. (2012). Online Gaming Motivations Scale: Development and Validation. In *Proceedings of the SIGCHI Conference on Human Factors in Computing Systems*, 2803–2806. http://dl.acm.org/citation.cfm?id=2208681

12 Rethinking UX in an Artificially Intelligent Future

George Mathew

INTRODUCTION

The technological landscape is undergoing a seismic transformation, driven by advancements in artificial intelligence (AI) and machine learning (ML). These technologies are not merely enhancing existing capabilities but redefining paradigms across industries. Integrating AI and ML into various domains has revolutionized user experience (UX) design.

AI's emergence as a dynamic and adaptive force requires rethinking traditional UX practices. Historically, UX design relied on deterministic models with predefined user journeys, static goals, and interfaces. However, AI-driven systems, with their probabilistic, data-driven, and evolving nature, challenge these approaches.

This paper explores how UX design must evolve to harness AI's full potential, positioning it as a core driver of enhanced user interaction rather than a supplementary tool for efficiency. It advocates for a change in thinking to align UX with AI's dynamic, adaptive nature. Drawing on academic studies and real-world examples, it examines challenges in traditional UX practices, the evolving role of designers, and the need for innovative approaches embracing uncertainty, adaptability, and human–AI collaboration. The proposed practitioner approach highlights the constructive interaction between ML and UX, envisioning a future where AI becomes UX itself.

LITERATURE REVIEW

The convergence of AI and UX design has been a subject of increasing scholarly interest. Verganti et al. (2020) highlight the transformative potential of AI in design innovation, emphasizing its role in fostering adaptive and personalized experiences. Similarly, Liao et al. (2020) propose frameworks for integrating AI into design processes, underscoring the need for collaboration between designers and AI practitioners. Affective Computing (AC) is a field of diverse disciplines that emerged over two decades ago, and among its aims is studying the interaction between humans

DOI: 10.1201/9781003557777-15

and machine technology (Calvo, D'Mello, Gratch, Kappas, 2014). Yang et al. (2016) argue for incorporating adaptive UI principles during the wireframing stage to anticipate the dynamic nature of AI systems.

Despite these advancements, challenges remain. Churchill et al. (2018) and Dove et al. (2017) identify significant gaps in designers' understanding of AI capabilities, which hinder effective integration. Main and Grierson (2020) discuss designers' attitudes toward AI as merely a creativity support tool only, revealing a spectrum of perspectives ranging from skepticism to optimism. All these studies from their vantage points collectively stress the urgency of reimagining UX design to align with the evolving technological landscape.

CHALLENGES IN TRADITIONAL UX DESIGN

Traditional UX design processes are predicated on deterministic models, where user journeys and system functionalities are meticulously mapped out in advance. While effective for static systems, these methodologies falter when applied to AI-driven interaction design, which is inherently adaptive and probabilistic. Three key challenges illustrate this disconnect:

1. Dynamic Behavior of AI Systems: Unlike rules-based software, AI outputs evolve based on user interactions. For instance, recommendation algorithms in streaming services adapt to users' habits, often diverging from initial design intentions (e.g., Spotify's evolving playlists).
2. Emergent Behaviors: AI-driven systems, such as chatbots, learn conversational patterns over time, resulting in responses that may deviate from predefined flows. If unaccounted for, this adaptability can lead to user confusion or dissatisfaction.
3. Personalization, Context, & Ethical Awareness: AI's ability to tailor experiences based on user context, such as personalized feeds, challenges the static nature of traditional designs. Users may struggle to understand or trust the logic behind dynamically curated content and responses.

THE NEED FOR A SHIFT IN UX DESIGN

These challenges necessitate a shift in UX design paradigms to embrace AI's fluidity and unpredictability. Even though many in the human–computer interaction (HCI) community promote AI as a design material toward creating new experiences, deliberate realignment to both strategy and execution is required in the design process. Embedded in this vision are the inherent interdependencies between the AI and the UX practices that UX needs the context awareness and ability to personalize, enabled by AI, and AI needs UX design to build trust and confidence in users. Ideally, AI and UX need to co-exist in a symbiotic relationship with one another, driven by the idea that "design defines products in people's minds."

Such integration of AI into UX processes would redefine the role of designers. Traditionally seen as architects of user journeys, designers must now become facilitators of dynamic human–AI interactions. This shift entails:

1. Understanding AI Principles: Designers must grasp the fundamentals of AI and ML, including their probabilistic nature and potential for emergent behaviors. This knowledge enables more effective collaboration with AI software engineers.
2. Anticipating User Needs: Adaptive design frameworks should account for diverse user interactions and potential outcomes. For example, incorporating implicit feedback loops can enhance personalization while maintaining user trust.
3. Prioritizing Explainability and Transparency: As AI-driven systems make complex decisions, designers must ensure that users can understand AI outcomes. Designing interfaces that explain AI reasoning fosters trust and accountability.

To address the challenges outlined, this paper proposes an approach for integrating AI into UX design. Key components include:

1. Dynamic Interaction Frameworks to replace static user flows with adaptable frameworks that evolve with user behavior, for example, conversational AI systems can leverage natural language processing to provide context-aware responses that generate conversational workflows.
2. Feedback Loops and Continuous Learning to embed mechanisms for collecting and analyzing user feedback, enabling iterative improvements both in AI behavior and UX.
3. Human–AI Collaboration through interface design facilitates seamless interaction between users and intelligent systems, emphasizing autonomy and trust to enhance user satisfaction and engagement.
4. Ethical Considerations that address ethical challenges, such as data privacy and algorithmic bias, by incorporating robust safeguards into the interaction design.

Several examples illustrate the practical implications of AI-driven UX design:

1. System Capability: Platforms like Netflix, TikTok, and Amazon demonstrate the potential of AI to create personalized UXs. However, their success hinges on transparent algorithms and user control over recommendations. For example, this author noticed that his TikTok reels were showing traditional Chinese remedies for a persistent cough, a symptom the author had at the time of writing.
2. Conversational AI: Virtual assistants such as Alexa, Siri & ChatGPT highlight the importance of adaptability and context-awareness. Designing these systems requires a balance between automation and user agency. For example, users filling in an online bank application should be answering questions about their situation rather than filling in forms.

3. Adaptive Interfaces: Mobile applications employing adaptive UI principles, as discussed by Yang et al. (2016), highlight the benefits of early integration of AI considerations in the design process. For example, many mobile apps regularly mine user behavior and context data to make personalized recommendations, predict travel time, availability of choices, and log behaviors like walking or sleeping.

These examples highlight the growing gap between the capabilities of AI systems and the constraints of traditional UX design. The deterministic, static outputs of the current process simply cannot keep pace with the fluid, adaptive nature of intelligent systems. A fundamental shift in our approach would recognize that AI is not merely a tool to be incorporated into existing UX frameworks, but is the driving force reshaping the very nature of UX.

A NEW APPROACH FOR UX DESIGNERS

1. The rise of AI accelerates the shift from product-centric to user-centric design. Rather than defining fixed features, UX designers must adopt a fluid process, leveraging AI's evolving responses to foster autonomy, trust, and explainability.
2. UX designers must transition from architects of fixed journeys to facilitators of human–AI interaction. This requires deep AI understanding, anticipation of emergent behaviors, and designing seamless human–AI collaboration with systems that can "think" and "do."
3. User interaction models should drive the software design process, ensuring the Software Development Life Cycle (SDLC) begins only after firming these models, as design defines how users perceive products.
4. Designers must adapt user stories for systems that learn and evolve. Repeated collaboration between UX designers and AI practitioners bridges gaps between design and development during the SDLC.
5. The focus shifts from precise user journeys to a collaborative process, enabling UX designers to revise journeys and stories through interaction testing, a new SDLC step for AI systems development.
6. Interaction design must ensure system explainability and address ethical safeguards on data privacy and algorithmic bias. Adding warnings, for example, for AI-generated content, human-in-the-loop mechanisms, and human control fosters trust.

CASE EXAMPLES FROM AUTHOR'S EXPERIENCES

For providing AI-powered diagnostic software for maintenance, repair, and overhaul of aircraft, ships, wind turbines, and other machinery and equipment, we rely on manuals, repair logs, and work instructions to provide recommendations on solving issues. Users utilize natural language queries to describe the problems they are trying to fix, and the system fetches them the relevant history or solution to help them quickly identify and repair the issues. The following examples illustrate the design considerations incorporated in our solutions to illustrate the UX design approaches described in this paper.

1. Relevancy Scores – Representing relevance as a percentage often confused users, who perceived low scores as inaccurate or unhelpful. Replacing the percentage with a color-coded bar (Figure 12.1) improved user perception, as it was visually intuitive and paired well with result rankings. Unlike rule-based systems that return no results if criteria are not fully met, AI systems offer probabilistic results even with minimal data, which users found valuable.
2. Dynamic Filters – AI-generated dynamic filters, shown in Figure 12.2, provide users with additional refinement options tied to their search query. Paired with weights, these filters offer insights users might not initially consider. Users can click on filter elements to trigger their follow-up queries, enabling deeper exploration and refinement of the results.
3. Combining Static and Dynamic Filters – Users valued starting with static filters (Figure 12.3) to narrow their query before entering a natural language search. After results were generated, dynamic filters provided further refinement, guiding users to the precise information they needed

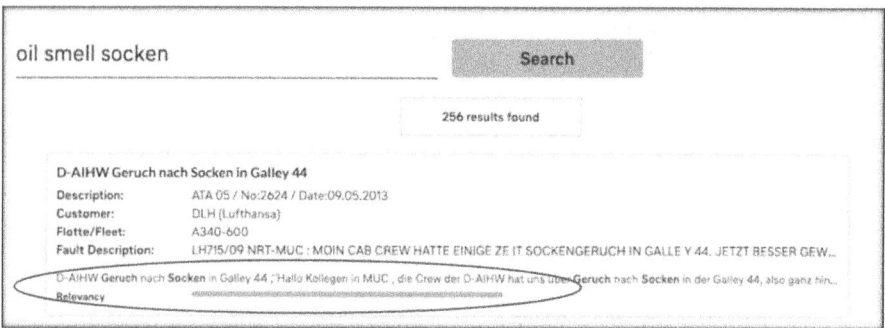

FIGURE 12.1   Color-coded relevancy bars were accepted more than percentages.

Image courtesy LexX Technologies

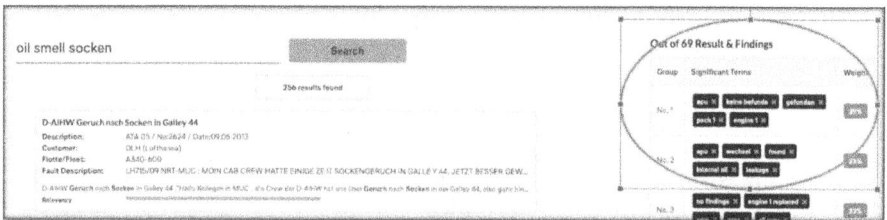

FIGURE 12.2   Dynamic filters accentuated the perception of assistance offered by the system.

Image courtesy LexX Technologies

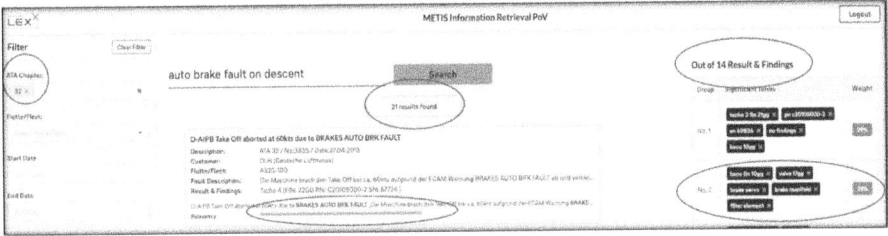

FIGURE 12.3 Choice and roles of static and dynamic filters enhance trust and explainability.

Image courtesy LexX Technologies

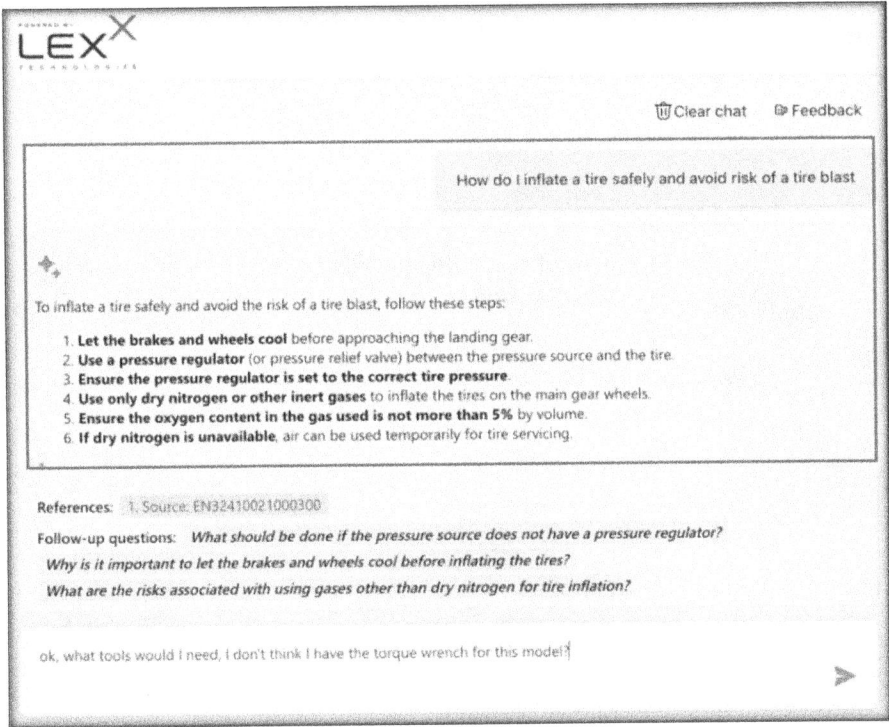

FIGURE 12.4 Conversational AI using generative content with sources referenced.

Image courtesy LexX Technologies

4. Generated Synopsis vs. Sections of a manual – Users preferred high-lighted sections of a source document (Figure 12.5) over generated content (Figure 12.4) due to their familiarity with exact references and training. However, educating users on how generated results synthesize multiple sources increased trust. Providing source access alongside generated content (Figure 12.6) further enhanced user confidence in dynamically curated results in conversational interfaces.

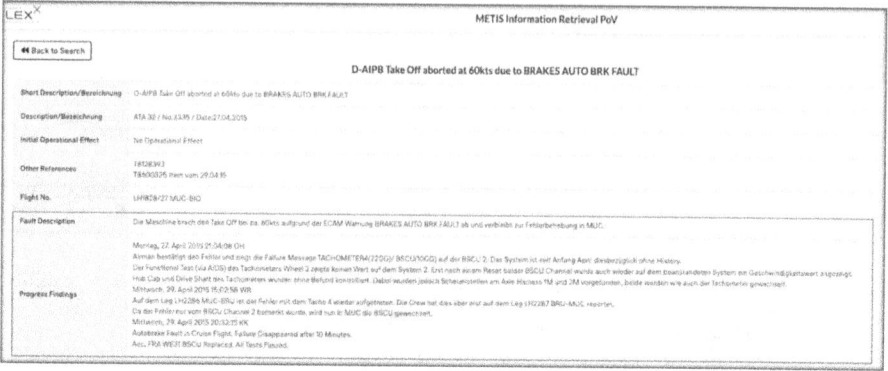

FIGURE 12.5   Responsive AI recommending sections of manuals without altering content.

Image courtesy LexX Technologies

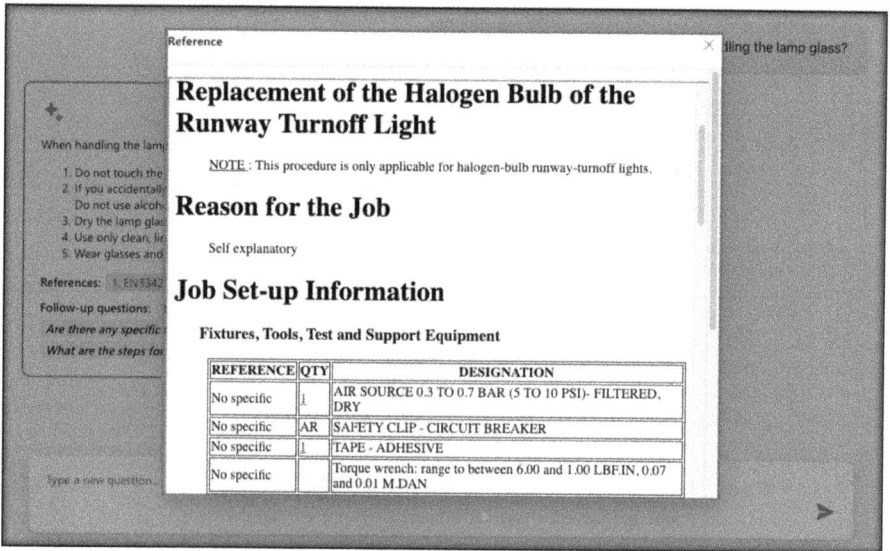

FIGURE 12.6   Citing source documents alongside generated content to enhance trust and explainability.

Image courtesy LexX Technologies

5. Combining Recommendations with Prompts – Users had mixed reactions to prompts over filters. While prompts encouraged new ways of interacting, users struggled to see the system as conversational. They expected consistent answers to similar queries, unlike in human Q&A interactions, where rephrasing is common. Adjusting prompt sensitivity and educating users

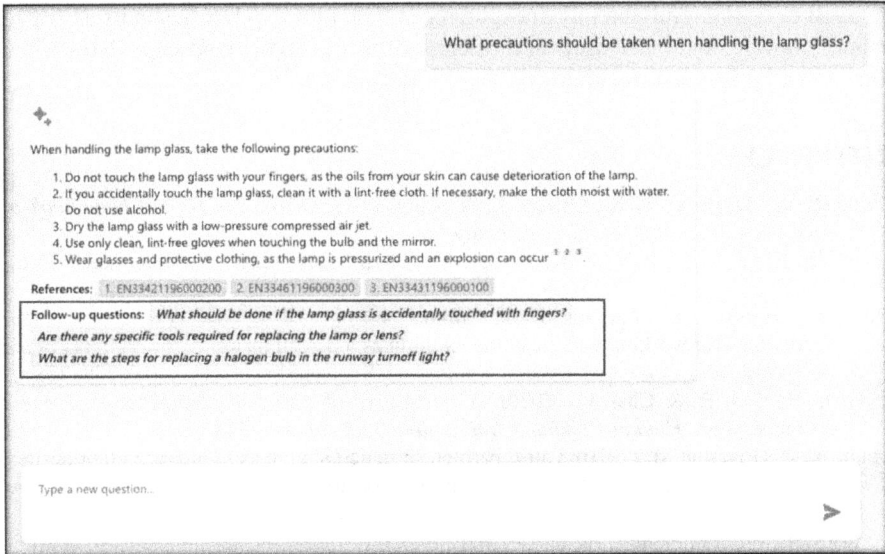

FIGURE 12.7 Combining generated content, citations, and recommending the next course of action increased confidence.

Image courtesy LexX Technologies

on the system's learning behavior improved engagement. When follow-up questions were recommended (Figure 12.7), users were more willing to explore and interact with the system until satisfied.

CONCLUSION

The integration of AI into UX design marks a paradigm shift, offering immense opportunities while presenting significant challenges. Designers must address uncertainties such as inference errors, unpredictable behaviors, and the time users need to adapt to intelligent systems. Overcoming these hurdles demands a cultural shift within organizations, emphasizing interdisciplinary collaboration among designers, data scientists, and AI practitioners, alongside a commitment to innovation and adaptability.

This paper outlines a clear path forward. By embracing dynamic interaction models, fostering human–AI collaboration, and addressing ethical concerns, UX designers can create intelligent systems that build trust, enhance satisfaction, and engage users in transformative ways. Traditional methodologies must evolve, reimagining the role of designers and adopting frameworks that align with the adaptive nature of AI-driven systems.

As AI reshapes the landscape of human–technology interaction, it is crucial to move beyond viewing AI as merely a tool for efficiency. Instead, organizations and practitioners must recognize it as a fundamental design material, integral to the creation of meaningful, intelligent experiences. The future of UX lies in its ability to adapt, innovate, and place users at the center of this evolving landscape. Now is the time to lead this transformation.

REFERENCES

Calvo, R. A., D'Mello, S. K., Gratch, J., & Kappas, A. (2014). *The Oxford Handbook of Affective Computing*. Oxford University Press.
Churchill, E. F., Allen, P. V., & Kuniavsky, M. (2018). Designing AI. *Interactions*, 25(6), 34–37.
Dove, G., Halskov, K., Forlizzi, J., & Zimmerman, J. (2017). UX design innovation: Challenges for working with machine learning as a design material. *Proceedings of the ACM SIGCHI Conference*, Denver, Colorado, USA from May 6–11, 2017, 278–288.
Liao, J., Hansen, P., & Chai, C. (2020). A framework of artificial intelligence-augmented design support. *Human-Computer Interaction*, 35(5–6), 511–544.
Main, A., & Grierson, M. (2020). Guru, partner, or pencil sharpener? Understanding designers' attitudes towards intelligent creativity support tools. *Computer Science*, arXiv abs/ (Jul 2020).
Verganti, R., Vendraminelli, L., & Iansiti, M. (2020). Innovation and design in the age of artificial intelligence. *Journal of Product Innovation Management*, 37(3), 212–227.
Yang, Q., Zimmerman, J., Steinfeld, A., & Tomasic, A. (2016). Planning adaptive mobile experiences when wireframing. *Proceedings of the ACM Conference on Designing Interactive Systems*, Brisbane, Queensland, Australia from June 4–8, 2016, 565–576.

Section 4

Making Design Truly Human and Planet Centred

Learning from Practitioners

13 Designing for Indigenous Communities

Alvin W. Yeo

PERSONAL JOURNEY AND MOTIVATION

Your career has focused on human–computer interaction and the implementation of Information and Communication Technology (ICT) in rural communities. What inspired you to pursue this path, and how have your experiences shaped your approach to designing for societal impact?

When I was completing my Bachelor of Computing and Mathematical Sciences degree, the paper that resonated with me was the course Human–Computer Interaction. The user interaction and thinking around user interfaces just made sense to me. For my PhD, I continued research in the Human-Computer Interaction area, specifically on how Culture affects software development.

It was through this interest and the research focus that I was introduced to the eBario project, a Bridging The Digital Divide (BTDD) project in Sarawak, Malaysian Borneo. The eBario project was a proof of concept to provide Internet access to a very remote area that lacked road access, a 24-hour power supply, and telecommunication infrastructure. While I initially joined to explore the field, I ended up working in that area for 16 years, gaining the privilege and opportunity to collaborate with indigenous communities in the remote and rural areas. Working on the eBario and later the eLamai projects gave me first-hand experience seeing how technology can bring about socio-economic impact, even in the most remote locations.

The research also helped me understand the challenges that rural communities face, inspiring me with the community's resolve to overcome their challenges. The research area introduced me to an interdisciplinary approach for working with and engaging indigenous communities.

DESIGN PHILOSOPHY

In your research, you've explored the reliability of usability assessment techniques in non-Western cultures. How do cultural contexts influence usability, and what considerations should designers keep in mind when creating solutions for diverse populations?

I used Hofstede's definition of culture, which defines culture as "the software of the mind", how a specific group of individuals with shared values and beliefs think and act. The distinction here is that culture is not just associated with international borders

DOI: 10.1201/9781003557777-17

169

or nationalities. Culture could also be the collective identity of those with shared interests, common ethnicities, or communities of people living in remote locations.

In order to address their needs and challenges, designers need to understand the users, their cultural contexts, bearing in mind that different people have different issues and face unique challenges. Different indigenous communities may have similar issues (e.g. trying to preserve their language and culture); however, each community may have distinct needs, facing different issues and obstacles that they may have to overcome.

At an abstract level, one size does not fit all. As such, approaches that may work for a specific culture need to be adapted to be effective in another. Taking the time to understand the community's needs can make all the difference in designing usable solutions.

e-BARIO PROJECT AND RURAL DEVELOPMENT

As the team leader of the e-Bario project, you worked on implementing ICT solutions in rural Malaysia. Could you discuss the participatory action research (PAR) methodologies employed and their effectiveness in engaging local communities?

While I led the team in later stages, my approach and thinking around working with communities was shaped by the social scientists who were initially leading the project. The PAR approach includes the process of collaboration in the research and action that are oriented towards social transformation. The communities are engaged at the very start of the project, first by obtaining their consent and agreement to participate in the project.

Similar to a human-centred approach, in PAR, which we employ, we place the community at the heart of the process of implementing ICT solutions. This involves understanding the problems faced by the community, the contexts within which they occur, as well as their goals and aspirations.

Solutions developed are also participatory in that they are co-created and the solutions are co-designed.

This approach helps to not only build trust with the community, but also helps to develop solutions that are targeted to their specific needs. Given the co-creation process, the community would "own" the solution, reducing the risk of the solution becoming a white elephant. A key component of the research team's work was the continuous reflection, self-improvement among the researchers.

ALTERNATIVE UX FRAMEWORK

Our book explores an alternative User Experience (UX) framework grounded in human values, needs, capabilities, and future impact. How have you incorporated these elements into your projects, particularly those focusing on BTDD in rural areas?

Since completing my project work in Malaysia, I have continued to conduct research with indigenous communities in New Zealand (see Yeo et al. (2022)). Upon reflection on the work that was carried out in Malaysia and now in New Zealand, the experiences are very similar. The research project work is very much led by the community's aspirations in keeping with their culture, their protocols, thinking, and values. For example, the Whakatōhea iwi in New Zealand are keen to participate in the digital economy and are open to exploring how technologies can be leveraged to help address their goals. Similarly, the aspirations of communities in Malaysia included a willingness to use technology to address issues such as preservation of culture. Grounding the work with the communities' aspirations provides the university with the opportunities to build a trusting, respectful, genuine, mutually beneficial, and long-term relationship with them.

In my opinion, building that trusting relationship is the key to success of the many projects carried out; deliver what you promise. The relationships will result in a continuous stream of projects, and the relationship is for life. The Malaysian relationship with the communities is still maintained by my ex-colleagues in Universiti Malaysia Sarawak (since 1998). The current work with the iwi has been going strong since 2016, with multiple projects being completed.

PARTICIPATORY AND INCLUSIVE DESIGN

Can you provide examples of how you've applied participatory design principles in your projects? What challenges and successes have you encountered in co-creating solutions with rural communities?

The iwi in New Zealand were interested in leveraging technology to preserve their Waiata songs (Yeo et al., 2022). The community guided on the elements which were important to them, not only the design interfaces, but also how the waiata were recorded and presented. Initially, challenges included understanding the vocabulary used, for example, the technical terms posed difficulties for the community; while cultural information required interpretation and understanding by the researchers. On this occasion, the content collection fell largely to one key individual who not only knew the waiata but also the composers of the waiata and how to gain access to them. The Waiata app is still in use by the community, and even by the elders.

ETHICS AND RESPONSIBILITY

How do you address ethical considerations in your design practice, especially when implementing technology in underserved communities?

Get informed consent in all activities, and if there are things you do not understand, ask. This works both ways and is essential for building a trusting relationship. Be clear on what happens to the data/content (in this case, the community retains ownership, control, and access of the data/content). Again, different communities may have different experiences and perceptions of how things work. Therefore, never assume what the community needs or wants. This ties to the guideline, kia tupato, to be careful.

FUTURE OF DESIGN

What emerging trends do you foresee in designing for societal impact, particularly in the context of ICT for rural development? How should designers prepare to address future challenges in this area?

In this case, this is for "ICT for indigenous communities". More communities are developing and are increasingly interested in preserving their culture. This is important for the communities in retaining their identity and the survival of the community. The indigenous knowledge may also be important to society, given that it may hold solutions to existing problems and sustainability.

Designers should be aware of these challenges, learn the language and culture of the target communities, to be able to better communicate, co-create, and co-develop better solutions. Designers should be aware of issues such as digital equity, data sovereignty.

ADVICE FOR ASPIRING DESIGNERS

What guidance would you offer to designers aspiring to create meaningful societal impact through their work, especially those interested in integrating human–computer interaction and participatory design in diverse cultural contexts?

Carrying out projects on leveraging technology for indigenous communities is fulfilling and rewarding. As the saying goes, "nobody can help everybody, but everyone can help someone". This is an opportunity for everyone to make a (small) difference to the world. That said, remember that we must stay humble, that we are co-creating with the community, rather than creating for the community, with the goal to ensure the community is not dependent on the researchers.

REFERENCE

Yeo, A., Hinze, A., Vanderschantz, N., Aporosa, S., Paruru, D. (2022). Mobile app development: Work-integrated learning collaborations with Māori and Fijian partners. *International Journal of Work-Integrated Learning*, 23(2), 237–258.

14 Design Led Futuring for the World

Anab Jain

PERSONAL JOURNEY AND MOTIVATION

Your work for the last few years has been focussed on your unique approach of design-led futuring that not only allows insight into forces at play but helps shape democratic, positive and rewarding futures.

What inspired you to pursue this path? Where did you start, and how did you get here?

I studied at NID, but I actually started with filmmaking. I was always interested in this kind of design. Later, I went to the Royal College of Art for my master's in interaction design. Eventually, with Anthony Dunne and Fiona Raby, it evolved into Design Interactions. I was still there at the time, working as a research fellow at the RCA.

My first research fellowship at the Helen Hamlyn Centre was focused on the future of work. That's when I really started exploring what design could do in relation to futures. Then came my time at Microsoft Research in Cambridge, where we began exploring ideas around the future of intelligence.

During that period, my partner and I started to realise that our experiences—through education and our various projects—had prepared us not only to create applications of emerging technology, but also to reflect on the implications of what we were designing. That became a turning point. I think that was the perspective that Anthony and Fiona were bringing into the mainstream conversation—one that invites critical engagement.

When you start thinking about technology, you inevitably run into the question: Who is it really for? Technology often gets framed as an answer, but the real question is—what was the question? Cedric Price asked that, and it's something we come back to often.

This reflection led to the founding of Superflux. It was around 2008/2009, just as the credit crisis hit. We were feeling its impact. It was disillusioning to see these vast, interconnected systems collapsing—and to realise how little we understood about their inner workings. It made us question the grand narratives shaping our world.

We realised then that we didn't want to just operate within those narratives. We wanted to ask—what lies on the margins? What alternative histories and futures are possible? That's what prompted us to start Superflux. From the beginning, our belief was: things don't have to be the way they are. We can imagine and make tangible other worlds—worlds that are more just, more humane, more diverse.

DOI: 10.1201/9781003557777-18

DESIGN PHILOSOPHY

In your work, you look at possibilities that mainstream design frameworks do not usually consider by critically questioning these grand narratives that we take for granted when working on a design project. Your design philosophy, in general, and design futuring philosophy, in particular, question these narratives and bring forth other ways to imagine alternate futures, in the here and now. I am very curious about the concept of being collectively hopeful and collective solidarity that is so central to your design futuring approach. This idea of collective solidarity is absent from our mainstream design frameworks.

Can you tell us more about that and how it fits into your futuring approach?

Jon and I bring backgrounds in film, art and design to the practice. And within design, it can be interaction design, experience design, critical or speculative design. But ultimately, the work we do is for people, and for the planet.

At Superflux, we imagine futures through storytelling, strategic foresight, speculative design and immersive experiences—spanning areas like climate, emerging technologies, AI, AGI, infrastructure design and more. For 16 years, we've guided businesses, governments and cultural organisations in thinking about the future from the future. This journey isn't just a straightforward path; it requires wayfaring—boldly stepping into future scenarios and engaging with the complex ecosystems around us. Such an approach allows people to engage emotionally with some of the challenges we face today—but more importantly, consider and imagine the vast potential of the future.

So whatever tools or skills we bring, the question we're asking is: How can the future be different from the one that's already being scripted for us? And that work—of imagining the future—is not something you do in isolation. It's a deeply collective activity.

I am aware of the privilege I have to be able to think about these things. Most people are too busy dealing with the present. So how do I use that privilege in a responsible way?

To me, solidarity is more important than empathy. Empathy can sometimes maintain a hierarchy—the designer who steps into someone else's shoes. But solidarity is about standing side by side, recognising that we are entangled in these challenges together. That we all have a stake.

And historically, the most powerful and transformative movements have been collective. We've seen that time and time again.

FUTURING THAT IMPACTS POLICY

Have you, in the course of your work, been able to impact those who make policies to see the future differently?

Yes, definitely. Most recently, we did a project called Ecological Intelligence, in collaboration with a policy lab in the United Kingdom and DEFRA—the Department for Environment, Food and Rural Affairs. We were asked to help policymakers think about the future of rivers and freshwater systems beyond 2040, since much of the current policy already looks ahead to that point.

We created a speculative piece where an AI speaks on behalf of a river. Instead of just advocating for human needs, the AI advocated for ecological systems. What if AI systems weren't just tools of human optimisation, but also represented the voice of ecosystems?

We presented this to policymakers. It landed well. The work has since been reused in other contexts. And I really hope it helped expand the frame through which environmental policy is viewed.

In another project, we brought samples of polluted 'future air' to a room of senior decision-makers, including the Prime Minister of the UAE, as part of their energy strategy discussions. Our question was: Is this the kind of air you want future generations to breathe?

It was provocative and unsettling—and that was the point. Did it change everything? Of course not. But they did invest significantly in renewables. And I do believe that embedding information within sensory, visceral experiences can have an impact. Because facts and data alone often fail to move us at the scale and urgency required.

NEW APPROACHES TO UX DESIGN AND RESEARCH

Since we are exploring new approaches to Design and Research, your idea of critical activism is very intriguing. There are some recent alternative design frameworks/approaches that are inspired by critical theory, such as feminist design or design justice. So when you talk about critical activism in the context of design, what does it mean?

I think when you talk about being a designer, what people expect is that you're shipping products in one form or another, whether that's services, or UX, clothing or tech products. But what we are doing through our work is trying to get people to think critically, to question. We often say that the work we are doing is designing questions.

And that's the critical activism. We are posing questions such as—what if the future could look like this? Or what if the present could look like this? We are not predicting the future because nobody can predict the future. Even with all the data, we cannot predict what the future will be.

So let's say that this job is about using creativity and imagination to get people to say,

> Oh, my God, okay. So what if in a particular future my apartment could be full of these things that are growing food, and only a part of my home is for actual living, and there will be no fridge, and there'll be no milk. And okay, so what?

So what is that getting them to do? That's getting them to think that

> what I've understood the future to be is not necessarily the only future that is possible. But hey, look! I didn't know you can build these tools with your hand. Oh, I didn't know we could grow food without any soil or water.

So the underlying social mission of the work we do is to spark critical thinking in people, because we think that when people start to think critically, then they start to question things. And when people start to question things, they demand things to happen differently, and that affects change. So it's not always that a solution has to be a tangible product. A solution can also be a question that makes you think and demand things to be different.

Hence, we are enabling and facilitating critical activism by designing the questions.

PARTICIPATORY AND INCLUSIVE DESIGN

Referring to your perspective of collective solidarity, in the course of your work, have you been able to facilitate or help community participation and creation happen?

We have been involved in, for instance, a project we did with young people, where we seeded the imagination that other worlds are possible. In other projects like Hope in the Heat, or one of the 1st projects I did called Power of 8, we were going out and talking to people and trying to create collectively what an optimistic vision of the future looks like.

But we've also done this kind of work in more sort of closed-room scenarios where you're working with leaders or decision makers who also sometimes are blinkered by a single idea of a future or a narrative that is supposed to be useful for quarterly returns. But it's not necessarily the best decision for the long term.

I want to also say that for facilitating community participation, I am inspired by history, like historically, communities that have self-organised and have imagined a different future and have collectively done the work they needed to do to have that different future. For instance, our entire freedom movement in India. Or if we talk about abolitionism, or about recent youth climate movements such as Fridays for Future. I am deeply inspired by people's movements.

ETHICS AND RESPONSIBILITY

You talk about rewilding the human spirit, and I view that as a unique ethical and responsible perspective towards a future for our planet that holds some hope.

Tell me more about this perspective.

I think that the actual words were from John, my partner, with whom I collaborate a lot. We were doing a research project with King's College, London, on a climate Futures Residency, and a lot of people were talking about rewilding. Rewilding is very important, but what we've heard a lot of has been talk about rewilding a part of the city, bringing back the hedgehogs and the otters and then let's just carry on, and it'll all be fine. But actually it won't be fine. And the point is that what we're trying to do is only fixing small gaps. They're not actually addressing the big problem that we have.

Because the deeper crisis is how we live. The structures we've built. The assumptions we've normalised. We really need to rethink how we live and how we do what we're doing. It has to change, and only when we are able to reorient ourselves to this urgency will we realise that this means the systems we have in place need to change fundamentally. All the sort of rewilding we do around our parks and sidewalks is just performative decoration.

And so we began thinking—what would it mean to rewild ourselves? We felt we needed to rewild our spirit, and we needed to rewild ourselves.

This is not just about 'greening' cities. It's about rethinking our habits, our desires and our relationships with the world around us. It's about shedding the idea that anything outside the current system must be worse.

Of course, this is such a scary thing to think of. People don't want to do it because there is an assumption that whatever else will emerge that is not what we have now will be terrible. That's not true. Take Gandhi's idea of swaraj—local self-reliance, handmade cloth and interdependence. It wasn't regressive. It was deeply radical. And people were content. That kind of rewilding—the rewilding of our inner compass— feels vital.

FUTURE OF DESIGN IN FUTURING

As you have often said, futuring using design is one way, but there are other ways of looking at the future. What do you feel is the unique contribution that design brings to the table versus very analytical ways of looking at the future? What is it about using design methods in futuring that creates a unique impact?

I think data is very important. I think scientific methods of modelling and generating information are also important because they give us an understanding of what's going on. However, as we are aware, Merleau-Ponty and other philosophers over many centuries have talked about the idea of humans not just thinking with our minds, but with our bodies.

How then do we act in the best interest of a future we cannot fully envision or feel? Embodied experiences provide a crucial answer by connecting us to the future through our bodies.

Mitigation of Shock (London, 2050), one of our experiential installations first presented in Barcelona's Centre de Cultura Contemporània, is one such example. At first glance, it seemed like an ordinary apartment. But step inside, and the future begins to reveal itself: indoor food production systems, newspapers, books such as 'Pets as Proteins' and 'How to Eat in Scarcity', local foraging maps, neighbours' growing notes and the overwhelming sense of damp air, heavy with the humidity of the rising seas. Every detail—from circuit board food computers to the radio transmission—spoke to a world adapting to climate disruption.

Following extensive research and prototyping, interviews with experts from NASA, the UK Met Office and Forum for the Future, we built an entire future London apartment adapted for living with the consequences of a climate catastrophe. Rather than sticking our heads in the sand or becoming overwhelmed with fear, we decided to catapult ourselves and others directly into a specific geographical and cultural context to experience the ripple effects of extreme weather conditions and resource scarcity.

Mitigation of Shock was intended to be more than an exhibition; it was a carefully crafted visceral encounter with a plausible reality of 2050: a future where survival demands ingenuity, resilience and adaptation. Our audiences didn't just contemplate what adaptation might look like—they began to see how they, too, could thrive in a radically altered world.

Human decision-making isn't driven by data alone. As cognitive scientist George Lakoff explains, emotions, metaphors, storytelling—and our deep desire to belong— are what truly shape our actions. With projects like the Mitigation of Shock, we explore the idea that stepping into an environment embodying these predictions, while still recognisable as a home, can speak directly to our sensory and cognitive apparatus. Using physical objects, sounds and smells, our work engages the senses we've evolved to rely on.

When we step into an imagined (future) world, it invites us to think differently about our role within it. When we inhabit potential futures—physically and emotionally—we begin to perceive, interpret and act on emerging possibilities, turning uncertainty into a space of opportunity.

ADVICE FOR ASPIRING DESIGN ENTREPRENEURS

When you first thought of creating super flux, did you find it difficult? Was it a difficult journey? What would you advise aspiring design entrepreneurs?

Yes, it was difficult and it still is. I think one of the things to do is to not be worried about size and scale and instead tread lightly. So if you, as an aspiring entrepreneur, start by saying, 'I'm going to have a boutique studio, I'm going to have 50 employees, and this is our offering.' Then you've already put yourself into a trap. You've created your own trap.

But if you are open and say, 'okay, I'll stay a little lightweight. I will collaborate with interesting people. I'll see where this goes,' you will be able to navigate the entrepreneurial journey better. If you have a capacity to do that, if you have a capacity to acknowledge that, that means embracing frugality, then it's possible. John and I lived in a shared house for a long time, doing what we wanted to do. We didn't have a lot of things, and we did not have the means to do much more. But we were quite determined that we wanted to push the creative agenda further.

Of course, I think it is not the same anymore. It is much more difficult for young people because they are in a much worse position than I was at that age. So I genuinely feel that young people do need support and encouragement to be able to try things.

One of the things that I think I'm taking away from this part of the conversation is that if as an aspiring design entrepreneur and you have a vision of what would be fulfilling for you, then it is ok to redefine what success would look like for you. You've got to be clear that there are things that you will not perhaps have, like your peers. You've got to be clear about what is driving you towards your vision and stay focussed on that.

Yes. You know, I speak to many young people, and they talk about constraints. So a lot of my conversations have been 'okay, let's look at your constraints and then see where are the little holes within your constraints that can be opportunities. Where are the cracks?' Maybe look at doing just one project first, one collaboration with someone. Just see how it goes.

I can see that redefining the journey of a design entrepreneur is, in itself, an important speculative design project, and you are one of the pioneers in redefining the traditional journey of a design entrepreneur.

And finally, is there a quote you would like to end the interview with?

Modernism's methodologies of mapping, designing, planning, for controlling and changing deeply complex systems may not be the answer to the challenges we face. Maybe we need to go underground — working in networked, symbiotic companionships, like mycelial arrangements, to generate infinite micro-revolutions.

https://medium.com/@anabjain/calling- for- a- more- than- human- politics-f558b57983e6

15 Design in the Healthcare and Pharma Industry in the United Kingdom

Elisa del Galdo

PERSONAL JOURNEY AND MOTIVATION

With over 30 years in User Experience (UX) research and design, spanning roles from practitioner to director, what initially drew you to this field, and how has your journey influenced your approach to designing for societal impact?

When I arrived at the Florida Institute of Technology, where I had chosen to study Marine Biology, I was already interested in psychology, engineering, science, research, and design. At the time, I didn't know what Human Factors (HF) was, but it was purely good fortune that my roommate's father was a man called Howard Harper, and the Senior HF Engineer at Sikorsky Aircraft. It was Howard who seeded my interest in HF, which I later went on to study in a Master's programme at Virginia Tech.

From my very first step into HF back in Florida, it has come down to showing interest, asking the right questions, and making connections in the industry. This is also good advice for how to carry out research and design.

On finishing my Master's, I went on to work with Digital Equipment Corporation (DEC) in the United Kingdom, in the International Products Group. I experienced first-hand the impact of design on users. Particularly, those who did not speak English and were unable to take full advantage of digital products that hadn't been designed for translation and localization. It's easy to think about what you need, but harder to think about what others might need, and that's where the skills and benefits of good UX lie.

DESIGN PHILOSOPHY

In your tenure as Head of Customer Experience at Blue Latitude Health, you emphasized the importance of customer-centric solutions in the pharmaceutical industry. How do you balance regulatory constraints with the need for innovative, user-centred designs in healthcare?

Balancing regulations with the need for user-centred designs is not as constraining as you may think. Yes, there are lots of rules and processes to follow as well as certifications that are necessary, but once you learn them, they are not as onerous

DOI: 10.1201/9781003557777-19

as expected. I always remind myself, the regulations are there for good reason, to protect vulnerable research participants and ensure that healthcare professionals are not unduly influenced.

Having said that, what can make research more difficult is the requirement for a research protocol to contain every question you are likely to ask in an interview. There is no spontaneous probing allowed in a research session. Every protocol must be reviewed and approved internally, which takes considerable time. This can be frustrating when you are in a research session and you want to probe deeper, but have not included the required questions in your protocol, stopping you from collecting those nuggets of insight that can drive innovation. This specific regulation takes away the spontaneous skill of a flowing research conversation and forces you to formalize and document every eventuality, but with good planning and forethought, you can reduce the likelihood of encountering any issues.

INTERNATIONALIZATION AND CULTURAL SENSITIVITY

As co-editor of 'International User Interfaces,' you've addressed the challenges of designing for diverse cultures. What strategies do you recommend for ensuring cultural sensitivity and inclusivity in global UX design projects?

SHOW INTEREST AND ASK THE RIGHT QUESTIONS

Designing for multiple cultures and markets can be a huge undertaking, especially if you are designing for diverse markets. Knowledge of a culture is always useful, as is a thorough literature search of related subjects. Be aware that culture constantly changes, so what you think you know, or what was reported by another researcher, may no longer be applicable in the present.

Asking the right questions is essential to ensure you cover all your bases. That means doing both qualitative and quantitative research to get a fuller picture of culture and needs across a larger portion of your target markets. Doing global design research using qualitative research first, on three of your largest and most diverse markets, is a good way to begin to understand the differences and similarities between markets. Be aware that countries that speak the same language or are geographically close may have very different cultures or needs. Countries will also have their own set of regulations and laws that you will need to understand and adhere to.

For qualitative research, always use an interviewer who speaks the local language and is also a UX practitioner. Much can get lost in translation, and their UX knowledge will mean they recognize statements that warrant further probing in line with the goals of your research. Data and insights from qualitative research are then used to inform a quantitative survey (questions and possible answers) for all other markets. This helps to ensure the questions and replies available are relevant and inclusive.

MAKE CONNECTIONS IN THE INDUSTRY

Having a large and culturally diverse team with individuals from your target markets is a benefit, but often not a common luxury. Often, research teams will recruit interviewers abroad via an agency. I found this, in most cases, to not be suitable, as they usually provide market researchers who do not have the UX expertise required. The best alternative is to create a global network of trusted UX practitioners, preferably with experience in your target industry. This is something I started early in my career and have benefited from. Working in internationalization, I have met and subsequently worked with practitioners from all over the world, making large international studies feasible, however, this type of network will take time to build.

ALTERNATIVE UX FRAMEWORK

Our book explores an alternative UX framework grounded in human values, needs, capabilities, and future impact. How have you integrated these elements into your projects, particularly those aimed at enhancing patient and healthcare provider experiences?

The needs, values, and capabilities of users you might work with in the Pharma space can be drastically different from what you would normally experience in other industries. You could be carrying out research for a new treatment protocol for patients with dementia, the terminally ill, or children. Their outlook and needs can be very different depending on what their condition is and how they approach their own condition.

You will find that patients are usually happy to talk about their experiences and illness with a researcher, even with something serious like cancer. This could be because they feel they can't talk to family or friends about their experience. There might be a fear of burdening others or creating a depressive mood with the topic of their condition. In this case, an interviewer who is willing to listen, ask relevant questions, and pay them is welcomed. As an interviewer, you must be receptive to hearing everything the participants want to tell you, whether relevant to your research goals or not. This might extend the time for the interview, but this process is as much about letting the participant feel safe to talk as it is about getting answers to your questions.

As well as taking care of your research participants, it's important to take care of yourself and your fellow researchers when interviewing healthcare professionals and patients. It can get emotional for all involved when discussing patient or practitioner experiences of a condition or treatment. Ensure that your researchers have time to 're-group' between interviews. Empathy and patience are key to creating a rapport and getting a deep understanding of your participants' experiences.

PERSUASIVE DESIGN IN HEALTHCARE

In your article 'Persuasion in Design,' you discuss the role of persuasive design elements. Can you share examples of how persuasive design has been effectively utilized to promote positive health behaviours?

A great example of this is when, at Blue Latitude Health, we found that terminally ill men are reluctant to talk about both emotional and practical aspects of death. To address this, we created a free online and offline tool delivered to metastatic prostate cancer patients and their caregivers at the point of diagnosis. The final patient and caregivers' supporting assets included a set of flash deck cards, a website, and multichannel patient and caregivers' materials. This utilized the persuasive design technique, Power of Free – we are more likely to use something if it's free.

The tool itself was a set of cards divided into four categories: practical advice (finances, employment, and travel), emotional support, treatment and reminders, and diet and exercise. Each card had a quote reflecting a real patient experience, intended to emotionally resonate with the patient or caregiver. The other side of the card showed where to get more information on the topic and advice on how to approach the topic with loved ones or health care practitioners (HCPs). This utilized Social Proof – We look to others to guide our behaviour by using quotes from other patients like them. The tool helped patients to initiate difficult emotional and practical conversations that were typically avoided, also connecting them to available support and encouraging discussions with HCPs, partners, family, and caregivers. Liking may also come into play as the people the patients use the cards with are those that are close to them and those they may want to please. To ensure trust in the content, the product received endorsement from the UK Oncology Nursing Society. Key charities and non-profits provided reputable content, support, and channels for distribution. This utilized the persuasive design technique of Authority – We look to experts in our decision making.

ETHICS AND RESPONSIBILITY

Working within the highly regulated pharmaceutical sector presents unique ethical challenges. How do you navigate these while ensuring that user needs and societal impacts remain central to the design process?

It is unfortunate that regulations can restrict UX best practices. In some cases, performing a normal UX iterative research and design approach can be near impossible. Your research protocol must go through an approval process that can take up to five weeks. If your participants are vulnerable patients, it can take even longer depending on their therapeutic area. To address this, you need to plan ahead and prepare your research materials will in advance of when you want to actually carry out your user research or user testing.

An additional regulatory barrier to overcome is that you can only ask questions that have been approved, which means you cannot ask any questions other than those that are on your approved protocol. Therefore, there are no opportunities for probing questions to clarify answers 'in the moment,' unless you have actually provided those questions in your approved protocol document. This results in protocols being 10s of pages in length just to ensure that you can get the most out of each research session. It can also make them slightly unwieldy during interview sessions. Again, planning ahead is your greatest strength in dealing with regulations in the Pharma industry.

Sometimes navigating the restrictions and regulations seems onerous, but it is worth spending the time up front on your protocol, as each participant in healthcare is precious and difficult to replace.

FUTURE OF UX IN HEALTHCARE

What emerging trends do you foresee in UX design within the healthcare industry, and how can designers prepare to address future societal challenges in this domain?

For over two decades, the transformation in the healthcare industry has been astonishing. Some good and some not so good. There are some amazing advances in healthcare, drug development, content for both the professional and layperson, and apps to diagnose, intervene, cure, and manage conditions. But there is also a lot available that may not be as useful as advertised. Take, for example, the thousands of mental health apps available to the public. These are not medical devices, so there is no licensing or verification of their ability to deliver on promised outcomes. How is a user (patient or practitioner) to know whether an application can deliver as promised or if it could potentially do more harm than good? How do people know which information provider to trust? How easy is the content to digest by a layperson? Does having access to content that cannot be fully understood do more harm than good?

A few areas come to mind when considering societal changes in the healthcare domain, which I have listed below.

a. Working on the development of services and tools that employ Artificial Intelligence (AI) and machine learning to ensure that they are employed for what they are good at and that the 'human touch' that is vital for medical treatment is not sidelined. Ensure that designs do not isolate patients from healthcare professionals, but support their interaction.

b. As populations become more health literate, the demand for apps and services that can demonstrate their efficacy will become greater. We will need to make sure our designs deliver in relation to health improvements and not just likability.

c. There will be a greater desire from patients to have access to healthcare content that is verified, easily digestible for a layperson, and also provides valuable information that is actionable.

d. I believe we are seeing a change in the value of UX roles due to the introduction of AI tools for creating applications, frontend design, creating databases, and even backend coding. The age of the designer as the 'gatekeeper' to UX is going, as these tools will provide the 'what' of design. Where a UX practitioner will continue and grow value is in research. Determining the 'why' that is provided by the human element of research to understand behaviour and inform solutions and drive innovation.

e. Last, but certainly not least, are digital therapeutics, evidence-based, software-driven interventions designed to prevent, manage, or treat medical conditions. Included are apps and services that encourage behaviour change to reduce the impact of behaviour-induced conditions, such as type 2 diabetes, drug and alcohol abuse, obesity, and smoking.

ADVICE FOR ASPIRING DESIGNERS

What guidance would you offer to designers aiming to create meaningful societal impact, especially those interested in the intersection of UX design and healthcare?

User experience (HF, ergonomics of research and design) is an amazing profession. It has allowed me to work across different industries, cultures, and geographies. This profession can keep people safe, make things more efficient and effective, more accessible, desirable (without having to artificially create a desire), see the world from a different perspective, understand someone's work and facilitate it, change behaviour for the better, and help to deliver desired outcomes. It doesn't get much better than that.

UX work in healthcare is interesting, rewarding, can be emotional, and requires a great deal of knowledge, alongside UX skills and experience. The 'ramping up' period on a project can be considerable and require reading medical research papers, understanding a condition, and being highly familiar with current and in-pipeline treatments. UX work in healthcare ranges across a very diverse landscape – pharmaceuticals, hospital design, digital therapeutics, training, unlicensed apps, biomedical equipment, medical devices, and service provider infrastructure, process, and delivery of healthcare service systems. I am sure I have missed a few, but that should provide an idea of the breadth and depth in the industry. If you don't like science, then this is not an area for you. If you do like science, then there is no better industry to ply your UX skills.

16 Culturally Relevant and Inclusive Design in Egypt

Ghada Refaat El Said

PERSONAL JOURNEY AND MOTIVATION

With over a decade of academic teaching and research experience in Information Technology and Management Information Systems, what inspired you to specialize in Human–Computer Interaction (HCI) and systems usability? How has your journey influenced your approach to designing for societal impact?

My journey into HCI and systems usability began with firsthand exposure to the challenges faced by users in underserved communities. Early in my career, I worked as a programmer and provided technical support for systems installed in schools located in rural areas of Egypt. Many of the users I assisted had humble educational backgrounds and were unfamiliar with digital interfaces. I observed their struggles as they attempted to navigate systems that were poorly designed, lacking user experience studies and proper usability testing.

At first, the common assumption was that the issue lay with the users and that their difficulty stemmed from a lack of digital literacy or training. However, as I continued troubleshooting and witnessing recurring patterns of frustration, I began to suspect that the core problem was not the users but rather the systems themselves. The interfaces were unintuitive, the workflows cumbersome, and the design failed to accommodate the needs and expectations of users.

This realization was solidified when I pursued my MSc in HCI at Nottingham University. There, I was introduced to the methodologies and research that validated what I had observed in the field: poor usability is not a user problem, but a design problem. I learned how usability testing, accessibility considerations, and user-centered design principles could transform how people interact with technology, ensuring that systems are empowering rather than obstructive.

Given that HCI and usability research remain underrepresented in Arabic-speaking countries, my concern has been to bridge this gap by advocating for culturally relevant usability studies. The absence of Arabic-focused research has resulted in systems that are designed with little consideration for linguistic and cognitive differences, making digital tools unnecessarily difficult for Arabic-speaking users. Through my work, I strive to elevate awareness about the significance of usability research and foster the development of systems that truly accommodate diverse user groups, ensuring that technology serves people effectively rather than alienating them.

DOI: 10.1201/9781003557777-20

DESIGN PHILOSOPHY

Your research includes a phenomenological study on metaverse-based learning opportunities and challenges. How do you envision the role of emerging technologies like the metaverse in shaping the future of education, particularly in developing countries?

This particular research, which takes a phenomenological approach to exploring metaverse-based learning, has given me a deep appreciation for both the promise and the practical challenges of integrating emerging technologies into education, especially in developing countries. I believe the metaverse has the potential to revolutionize education by offering immersive, interactive, and personalized learning experiences that go far beyond traditional classroom settings. For students in developing regions, this could mean access to virtual science labs, historical simulations, or even real-time collaboration with peers and educators from around the world, opportunities that are often limited by local infrastructure or resources.

However, the study also highlights the very real barriers that must be addressed. Many communities still face limited access to high-speed internet, reliable electricity, and affordable digital devices. These issues can deepen the digital divide if not proactively managed. Moreover, there are concerns around data privacy, digital literacy, and the cultural relevance of educational content delivered through the metaverse. To truly benefit learners in developing countries, we need inclusive strategies such as public-private partnerships to subsidize access, the development of localized and multilingual content, and training programs that build digital skills among both students and educators.

In essence, I see the metaverse not as a replacement for traditional education but as a powerful supplement, one that, if implemented thoughtfully and equitably, can help democratize access to quality learning and empower communities that have historically been underserved.

CULTURAL CONTEXT IN DESIGN

In your work on e-commerce adoption, you've explored the role of culture in technology use. What strategies do you recommend for designers to ensure cultural sensitivity and inclusivity when creating digital solutions for diverse populations?

Cultural context is not just a background variable; it plays a central role in how users perceive and interact with e-commerce. For designers aiming to create culturally sensitive and inclusive digital solutions, especially in high uncertainty avoidance cultures like Egypt and many other Arab countries, trust-building mechanisms should be a priority, emphasizing transparency, recognizable branding, and consistent user experiences to reduce perceived risk and build confidence.

Localization must go beyond translation. Cultural nuances such as preferences for one-to-one interaction, concerns about online payment security, and the importance of community validation should inform interface design and user flows. Involving local users in the design process is essential, designing with—not just for—target communities is the key.

ALTERNATIVE UX FRAMEWORK

Our book explores an alternative UX framework grounded in human values, needs, capabilities, and future impact. How have you integrated these elements into your projects, especially those aimed at enhancing technology adoption in developing regions?

Integrating human values, needs, capabilities, and long-term impact into UX design is not only essential but transformative. Rather than treating users as passive recipients of technology, view them as active participants with unique cultural, social, and economic contexts.

Human values such as trust and empowerment should be prioritized. In many developing regions, users may approach digital platforms with skepticism due to past experiences or systemic barriers. Designing with transparency, respect for privacy, and clear communication helps foster trust and a sense of agency. Also, we need to focus on real human needs, which often go beyond functionality. For instance, in areas with limited infrastructure, users may need offline access, low-bandwidth interfaces, or multilingual support. Understanding these needs requires deep engagement with local communities and a willingness to adapt solutions to their lived realities. Design interfaces that are intuitive, forgiving, and supportive, often incorporating visual cues, guided interactions, and culturally familiar metaphors to reduce cognitive load and build confidence.

PARTICIPATORY AND INCLUSIVE DESIGN

Can you share examples of how you've applied participatory design principles in your research? What challenges and successes have you encountered in co-creating solutions with local communities?

I participated with an NGO to co-design a mobile health (mHealth) app with local health workers and patients. We conducted community workshops with local participants where users helped define features, language, and workflows. Participants shared their experiences, challenges, and aspirations related to mHealth app use. These sessions helped us uncover insights, such as low tolerance for ambiguity and uncertainty while using mHealth apps as we discovered that receiving health advice from a mobile application presents numerous risks for users; the preference to receive medical terms in user's natural language with high level of explainability; and the preference to, at some points, receive a human-to-human interaction with a healthcare provider, with a preference of provider gender, especially for female users who showed preference to interact with female doctors.

These workshops were followed by iterative Testing as usability testing sessions were conducted with the same participants at multiple stages, ensuring the design evolved with their input. The process wasn't without challenges. One of the main difficulties was managing expectations as participants often hoped for immediate solutions or tangible outcomes, while our design process was iterative and exploratory. There were also logistical hurdles, such as varying levels of digital literacy, and ensuring inclusive participation across gender and age groups.

ETHICS AND RESPONSIBILITY

Considering your research on mobile health applications, how do you address ethical considerations, such as user trust and data privacy, in the design and implementation of mHealth solutions?

You must be referring to the research: *"Factors Affecting mHealth Technology Adoption in Developing Countries: The Case of Egypt"*. Well, ethical considerations such as user trust and data privacy are central to the successful design and adoption of mHealth solutions. My research revealed that trust, perceived risk, and governance are among the most critical factors influencing users' willingness to adopt mHealth apps.

To address user trust, the research emphasized the importance of perceived reputation and transparency. Users are more likely to trust platforms that are endorsed by reputable healthcare providers or certified by government authorities. Therefore, mHealth apps should clearly communicate their affiliations, data handling policies, and service quality standards. Users in the study were more comfortable using apps that were governed or regulated by public health institutions rather than private entities alone.

Regarding data privacy, the study found that users are highly concerned about the misuse of personal health information. Participants expressed a strong desire for clear privacy policies, control over their data, and assurances of confidentiality. Explainability was also found to be very important; this is the ability of the app to clearly communicate how recommendations are made and how data is used. This is especially important in healthcare, where decisions can have serious consequences.

FUTURE OF DESIGN

What emerging trends do you foresee in the field of HCI and systems usability, particularly concerning societal impact in developing countries? How can designers prepare to address these future challenges effectively?

The field of HCI is evolving rapidly, and several emerging trends are poised to significantly shape its societal impact in developing countries. One of the most prominent trends is the localization of AI and intelligent systems. As AI becomes more embedded in everyday technologies, there is a growing need to ensure that these systems are culturally aware, linguistically inclusive, and ethically aligned with local values. This means designing not just for usability, but for cultural relevance and social equity.

We're also seeing a growing emphasis on participatory and community-driven design. As digital technologies become more integrated into public services like health, education, and finance, it's crucial that local communities are not just users but co-creators. This ensures that systems are grounded in real needs and foster long-term engagement and trust.

ADVICE FOR ASPIRING DESIGNERS

What guidance would you offer to designers aiming to create meaningful societal impact through their work, especially those interested in integrating HCI principles and focusing on technology adoption in diverse cultural contexts?

Designers must cultivate interdisciplinary skills combining technical expertise with cultural sensitivity, ethical reasoning, and systems thinking. They should also embrace inclusive research practices, such as ethnographic fieldwork, co-design workshops, and iterative testing with diverse user groups. Moreover, staying informed about policy, data governance, and digital rights will be essential as HCI increasingly intersects with public infrastructure and civic life.

17 Applying HCD, Design Thinking, and Ethnography to Codesign Value Propositions for Patient Experience in Spain

Joan Vinyets i Rejón

PERSONAL JOURNEY AND MOTIVATION

Your career spans anthropology, design, and innovation, with a focus on human-centered approaches. What inspired you to integrate these disciplines, and how has this multidisciplinary perspective influenced your approach to designing for societal impact?

Both disciplines are human-centered and inherently complementary. Anthropology, through ethnographic methods such as immersive fieldwork, user narratives, in-depth interviews, and contextual observations, facilitates a deep and nuanced understanding of human experiences. Design thinking, by contrast, offers a structured approach to translating these insights into actionable outcomes. It systematically integrates user needs and feedback throughout the creative development process, fosters empathy, promotes collaboration within interdisciplinary teams, and employs iterative, rapid prototyping to explore and refine potential solutions while mitigating risk.

DESIGN PHILOSOPHY

In your work, you emphasize the importance of human-centered design and applied anthropology. How do these principles guide your projects, particularly in creating value-driven solutions that address societal challenges?

Contemporary societal challenges and the interventions designed to address them are often highly complex, requiring multidisciplinary expertise and complementary approaches. In this context, drawing upon recent work, it is evident that anthropology

DOI: 10.1201/9781003557777-21

and design thinking can serve as mutually reinforcing frameworks, both grounded in a human-centered perspective.

Anthropology offers an immersive, empathetic lens through which to gain a deeper understanding of people's perspectives, behaviors, and mindsets. This ethnographic approach facilitates rich, contextual learning that is essential for meaningful insight. In parallel, design thinking introduces a structured, iterative, and collaborative process for creative problem-solving. It emphasizes empathy, co-creation, and rapid prototyping, enabling teams to explore, test, and refine solutions with a focus on user needs and behavior.

Together, these approaches support the sustainable adoption of solutions, particularly in contexts requiring behavior change and improved user engagement—such as overcoming barriers to access and increasing the effectiveness of social interventions and programs.

However, many social projects risk developing solutions without meaningful user involvement. Here, the integration of anthropological inquiry and design thinking becomes especially valuable. By generating early user insights and facilitating experiential learning, this combined approach enables the continuous refinement of ideas based on what emerges from the users themselves. Rapid prototyping and iterative testing support a process of adaptive learning, allowing teams to pivot and enhance solutions efficiently within a collaborative, user-driven framework.

PATIENT EXPERIENCE AND HEALTHCARE INNOVATION

As the Head of Patient Experience at SJD Barcelona Children's Hospital, you've implemented methodologies like design thinking and ethnography to enhance patient care. Could you share specific examples of how these approaches have transformed patient and family experiences within the hospital?

Anthropology and design both support multidisciplinary approaches to problem-solving, enabling the emergence of novel perspectives. This is particularly valuable in healthcare, where such approaches can help Health Care Professionals (HCPs) perceive and interpret complex challenges differently and connect disparate insights across the care continuum. By integrating anthropological perspectives—especially through patient narratives—stronger human connections between professionals and patients have been fostered, leading to more empathetic care. This integration has also enhanced communication not only with patients and their families but also among members of interdisciplinary care teams.

One practical example of this integrated approach is the "First News" ("Primera Noticia") intervention, developed to bridge cultural and communicative gaps between physicians and patients. The project focused on improving HCP communication strategies, particularly in the emotionally charged moment when delivering a serious diagnosis to a child and their family. Ethnographic methods—including in-depth interviews, observation, and shadowing of patients, families, and healthcare providers—revealed that this moment of disclosure is a pivotal point in the patient journey. Termed a "moment of truth," it significantly shapes the subsequent experience of care for families.

The findings highlighted that HCPs often face these moments without adequate knowledge of the family's emotional disposition, communication style, or coping mechanisms. This increases the risk of miscommunication or unintentional emotional harm. In response, a training program and practical guide were co-created by families, patients, and professionals who had previously undergone this experience. These resources aimed to equip HCPs with tools to better manage the complexity of delivering bad news with sensitivity and awareness.

A key insight from this work was the recognition of diverse relational and communicative styles among patients and families. Individuals vary in their pace of life—some being more reflective, others more action-oriented—and in how they make decisions, whether driven by rational analysis or emotional resonance. Consequently, effective communication must be tailored to these relational patterns. For instance, some recipients may prefer objective, data-driven explanations, while others respond better to metaphorical language and narrative framing. Recognizing and adapting to these differences is critical to minimizing emotional distress and enhancing the quality of patient-HCP interactions, especially during vulnerable periods.

The outcomes of the "First News" intervention have been promising, resulting in improved patient and family experiences during initial diagnosis and fostering a shift in the way HCPs approach such sensitive situations. The success of the intervention has further led to the development of a simulation-based training program that prepares professionals to navigate these high-stakes interactions with empathy and confidence.

ALTERNATIVE UX FRAMEWORK

Our book explores an alternative UX framework grounded in human values, needs, capabilities, and future impact. How have you incorporated these elements into your projects, especially those focusing on healthcare and social innovation?

As a patient experience team, we actively promote a value co-creation approach across the projects in which we are involved. Each initiative begins with a thorough identification of the core problem and a contextual understanding of the environment in which it occurs. Central to this process is the continuous engagement of patients affected by the issue—from initial data collection through co-design workshops, pilot testing, and outcome evaluation.

We employ a diverse set of participatory tools—ranging from creative and empathic methods to projective techniques—within a flexible and iterative framework that is responsive to evolving contexts and needs. Emphasizing the inclusion of plural perspectives, we adopt a facilitative role, maintaining openness and adaptability throughout the project lifecycle.

Our role as facilitators involves providing a dynamic methodological framework that supports: (1) assessing the situation from the perspectives of patients and caregivers directly experiencing the condition; (2) identifying user profiles and their underlying needs (the "why"); and (3) generating and refining potential solutions that are aligned with the actual challenges. This human-centered, participatory approach ensures that the solutions developed are grounded in real-world insights and have greater potential for adoption and impact.

CRITICAL RESEARCH AND THEORY

Your approach integrates methodologies from anthropology and design thinking. How have critical research approaches influenced your work, particularly in fostering insight, critique, and transformation in design processes?

Critical research emerged from the imperative to interrogate power structures, systemic inequities, and their broader societal implications, particularly in sustaining the status quo. In clinical settings, similar power dynamics and disparities are present—often rooted not only in hierarchical knowledge structures but also in cultural and social diversity. In this context, anthropology and design thinking play a crucial role by providing contextual understanding, nuance, and participatory processes that challenge and transcend the traditionally dominant techno-scientific paradigms that shape healthcare delivery. These disciplines encourage the integration of patient and caregiver perspectives early in the exploration and design phases, thereby increasing the likelihood that resulting solutions genuinely reflect user needs.

In our projects, we actively recognize and address disparities related to gender, ethnicity, sexual orientation, socioeconomic status, and other social determinants of health. Our approach goes beyond a narrow focus on disease, organ dysfunction, or symptoms; it seeks to understand the lived experience of illness within its broader sociocultural context. The outcomes of our work are not limited to the generation of knowledge or insights, but extend to the transformative impact of these insights when applied through co-creation. This transformation is reflected in shifts in perceptions and attitudes—both among patients and healthcare professionals—ultimately contributing to the delivery of more holistic, equitable, and human-centered care.

PARTICIPATORY AND INCLUSIVE DESIGN

Can you provide examples of how you've applied participatory design principles in your projects? What challenges and successes have you encountered in co-creating solutions with diverse communities?

In a recent project, we collaborated with the Obesity Unit to design a new care model for onboarding patients with obesity. This model emphasizes patient-centered care and enhanced coordination among various healthcare professionals, including endocrinologists, nutritionists, and psychologists. Through patient journey mapping, we documented and analyzed the full continuum of the patient experience—from the initial medical consultation to post-treatment follow-up—and shared these insights with the healthcare team. This process revealed a significant pain point.

The first interaction patients had with the onboarding process involved a focus on weight measurement. This weight-centric approach often led to feelings of humiliation and shame among patients, thereby negatively impacting their engagement with care. To address this issue, we prototyped alternative onboarding processes and developed patient experience interventions through co-creation with patients, families, and healthcare professionals.

Rather than centering the initial visit on weight, we proposed a partnership-based approach. In this revised model, healthcare professionals adopt a more holistic view during the first consultation, prioritizing the establishment of a collaborative relationship. This involves understanding the patient within their broader context—exploring their needs, preferences, and goals—and assessing whether they have the knowledge and support required to actively participate in their care. Early pilot testing of this model demonstrated improved patient adherence and reduced the stigma often associated with obesity in clinical settings.

ETHICS AND RESPONSIBILITY

In your publication "Un mundo en clave F," you discuss new dynamics of wealth creation based on values attributed to women. How do you address ethical considerations in your design practice, particularly when aiming for social sustainability and inclusivity?

A growing body of evidence, including numerous peer-reviewed studies and publications—as well as findings presented in my book—indicates that, on average, women tend to exhibit higher levels of empathy than men. Recognizing these gender-based differences in empathic capacity enables researchers to better understand how certain aspects of the patient experience disproportionately affect specific groups. This insight is critical for designing more responsive and inclusive healthcare interventions.

With regard to inclusivity, our co-creation practices are guided by a commitment to equity and fairness. We strive to ensure that the voices of all patient groups are respected and represented, adapting our methods and activities to meet diverse needs. This includes accommodating individuals with physical or cognitive impairments, as well as those from varied cultural and linguistic backgrounds. Additionally, we place a strong emphasis on safeguarding participants' privacy and ensuring robust data protection throughout the research and design process.

FUTURE OF DESIGN

What emerging trends do you foresee in design for societal impact, and how should designers prepare to address future challenges in areas like healthcare and community well-being?

There are several critical challenges in healthcare and wellbeing where design can play a transformative role. Key opportunity areas include:

- **Advancing healthcare sustainability through circular models**: Design can contribute to the transition from linear to circular healthcare systems by optimizing the use and flow of resources such as water, energy, metals, plastics, and chemicals. This includes strategies to minimize waste, reduce reliance on single-use products, and promote the safe and effective reuse of medical equipment at the end of its lifecycle.
- **Servitization of products, care spaces, and medical devices**: There is growing potential to reimagine care environments and technologies not only as isolated products but as integrated service-oriented systems. This

includes designing healing spaces, devices that support treatment adherence and patient education, and comprehensive solutions that enhance care continuity and the efficiency of healthcare delivery processes.

- **Enhancing inclusivity for diverse user groups**: Design must address the varied needs of different populations, including older adults, children and adolescents, individuals with cognitive and behavioral disorders, and people from diverse cultural and linguistic backgrounds. Inclusive design practices can ensure accessibility, cultural relevance, and psychological safety for all users.
- **Empowering patients, families, and caregivers**: Information design, visualization tools, and interactive resources can enhance communication between healthcare professionals and patients, enabling shared understanding and collaborative decision-making. Moreover, co-created narratives and training materials can equip patients and caregivers with the knowledge and skills needed to understand their condition and treatment, thereby improving engagement and adherence.
- **Supporting behavior change and long-term motivation**: Effective design can foster behavioral change by leveraging motivational principles and feedback mechanisms. By creating engaging, personalized, and psychologically attuned interventions, design can support users' intrinsic needs for autonomy, competence, and relatedness—key drivers in sustaining health-promoting behaviors throughout the illness experience and recovery process.

ADVICE FOR ASPIRING DESIGNERS

What guidance would you offer to designers aspiring to create meaningful societal impact through their work, especially those interested in integrating human-centered design and applied anthropology?

At the core of any societal challenge lies its context. Complex problems resist simplification, and therefore, one must avoid seeking prescriptive, one-size-fits-all solutions. Effective problem-solving begins with a deep understanding of the specific context in which the issue is embedded. It is important to acknowledge that no matter how thorough the analysis, it is rarely possible to obtain all the information needed to reduce a complex problem to a simple or definitive one.

Instead, a process of continuous experimentation, iteration, and learning is essential. The design journey should be viewed as an evolving continuum—one that generates insights, refines ideas, and incorporates key elements of user experience through co-creation. Crucially, it must be recognized that replication is inherently difficult: what proves effective in one context may be entirely inappropriate in another.

Lastly, it is imperative not to begin with a pre-defined product in mind. Let the solution emerge through the process, grounded in contextual understanding and shaped by the needs, experiences, and participation of those it is intended to serve.

18 Responsible Design

V. Sudhindra

PERSONAL JOURNEY AND MOTIVATION

Your career spans roles such as Founder & CEO of design. Unbounded, Chief Design Officer at IBM India, and Experience Design Director at SapientNitro. What inspired you to pursue a path in design, and how have these diverse experiences shaped your approach to creating societal impact through design?

I came into design quite by accident. In 1999, during a transitional phase in our family, I stumbled upon a new, expensive design course in Bangalore. I'd always had an artistic streak—dabbling in literature, writing, and reimagining the world around me. There was a legacy too—my great-grandfather was an artist-in-residence at the Mysore court and briefly Principal of JJ School of Arts. I didn't understand the weight of that heritage then, but it felt right.

That course ignited a discipline and rigor in me. Initially, I was focused on craft and output. But a colleague introduced me to Human-Centered Design—and that shifted everything. I moved from making things to enabling outcomes: framing problems, facilitating visioning, enabling agile collaboration, and co-creating futures.

At SapientNitro, I worked on digital transformations. At IBM, I embedded design into business strategy, helping shape a lasting design culture. My worldview has been shaped across geographies—India, Dubai, Singapore, and now Australia. Over time, I've come to see design not as fixing, but as facilitating—creating the conditions for systems and people to better understand each other.

DESIGN PHILOSOPHY

In your presentation "Future of Design," you discuss the evolution of design in the context of emerging technologies. How do you envision design as a bridge to address societal challenges in an increasingly connected and technologically advanced world?

Design today is an enabler, not just an output. In my talk *"Future of Design,"* I propose that design must sit at the epicenter of ecosystems—not as a finishing layer, but as a foundational force.

Take the EV innovation workshops I led in Australia. We weren't just exploring EV infrastructure—we were probing cultural habits, fears, and aspirations. What would it take for a community to embrace clean transport? That's where design becomes a bridge: not between device and user, but between intent and impact, policy and people.

DOI: 10.1201/9781003557777-22

DESIGN CHRONICLES AND STORYTELLING

As the author of "Design Chronicles," how do you perceive the role of storytelling in design? Can you share an example where narrative has been pivotal in driving social change through a design project?

Storytelling is what makes us uniquely human. And design, well, is unique to us too. A conscious choice made that creates a desired outcome is what is design. Stories help articulate and bring good design to life. And great design creates stories that last long in the fabric of space-time.

One powerful example is the *samShiksha* project, a Government of India initiative to democratize learning, provide alternative pathways to higher learning, and eventually build a platform for online degrees. In this case, narrative was central to driving educational equity. We began by deeply understanding real stories through personas—like Gaurav, a student struggling with English and outdated content, and Dr. Amrita, a passionate teacher limited by rigid curricula. Their experiences weren't just research artifacts; they became the emotional anchors of our design.

These narratives helped us design a culturally rooted, inclusive platform. We created personalized content pathways for students like Gaurav, community forums to bridge language gaps, and teacher tools to enable more relevant pedagogy. Even the design language—drawing from Alpana art—reflected Indian authenticity and belonging.

By centering the platform around these lived experiences, we moved beyond functionality to build a learning ecosystem that could genuinely empower both students and teachers. Through the use of narrative, we transformed what could have been a functional app into a movement—one that could restore agency, belonging, and opportunity for students and educators across India. It reminded me how powerful storytelling can be—not just to inspire design, but to inspire change.

ALTERNATIVE UX FRAMEWORK

Our book explores an alternative UX framework grounded in human values, needs, capabilities, and future impact. How have you integrated these elements into your design processes, particularly in projects aimed at enhancing social sustainability?

There is for sure a need for an alternative UX framework, especially in the business context of today. Design practices across the world are falling head over heels to showcase their "AI savviness" and have abandoned the very fundamentals that have kept us in good stead—value-based design, human-centered design, tech for good. So, I believe any framework "grounded" firstly in human needs, and then grounded in human values, is the need of the hour.

I have always tried to integrate value-based design in my professional life. I have presented across various industry forums various times on the topic of "Responsible Design," much before it became fashionable to do so.

In the State Bank of India (SBI) Credit Card redesign, we made a conscious choice to prioritize responsible usage over aggressive activation. We didn't just look at customer journeys; we mapped aspiration journeys. We asked: What does "credit" mean to a first-time user in a tier-3 town? How can a credit card support a lifestyle and not become a financial burden to the user? And these questions brought us insights to integrate human values within the business context.

One of the glaring examples was when I led an initiative for British American Tobacco (BAT). It's hard to bring in human values as it is in regular brands, but when the brand is Tobacco, it becomes doubly challenging. But we took up the challenge and made several designs celebrating human values within the context of their digital ecosystem. Incentives for daily exercises and healthy habits, smoking tracker to monitor and limit with rewards for reducing their daily intake were some of the game-changing ideas that we presented to the BAT leadership. It requires designers to take the lead and have these conversations even at the risk of short-term business loss.

That requires courage, reframing Key Performance Indicators (KPIs), and listening more than we ship.

PARTICIPATORY AND INCLUSIVE DESIGN

Can you provide insights into how you've applied participatory design principles in your work? What challenges have you encountered in co-creating solutions with diverse communities, and how have you addressed them?

Participatory design is at the heart of design thinking and social change. In all our design activities, I have strived to bring a diverse set of users—be it vulnerable communities, senior citizens, students, and other groups as appropriate to the problem we were solving for. It's enriching to co-create solutions with diverse communities. The kind of insights you learn is most often mind-blowing, and when applied toward solutions, they end up being exponentially better. In a recent project in Australia, our objective was to better higher financial well-being for members of a bank. We had the opportunity to study spending patterns of people living in metros vs regional areas, senior citizens, families living with their adult children in the same household vs downsizers, and vulnerable communities.

One of my most humbling experiences was at Jana Bank, the microfinance bank in India. In a remarkable workshop, we had the senior leadership of the Bank participate with the frontline sales staff. What we realized during that interaction was that it was their very first time coming together as a group in such a professional setting to solve some significant problems within the organization. There was one particular moment when the CEO spoke of the problem he was facing because of increased attrition from the frontline sales staff and how it affected his bottom line. That prompted a frontline staff member to get up and articulate the reasons for such high attrition. "It was long hours, no work life balance and compromised safety and well-being of the employees that caused them to leave the job early, even at the cost of no clarity of their own future," he said. As he said this, there was an awkward silence. Nobody had ever spoken to their CEO so candidly before. Then the ice broke. There was enlightenment on the issue on both sides.

That discomfort—that friction—was design doing its real job. Participatory design is messy. People cry. People disagree. But that's the price of inclusion. You must sit in discomfort long enough for real understanding to emerge.

ETHICS AND RESPONSIBILITY

With technology's pervasive role in society, how do you approach ethical considerations in design? What frameworks or practices do you employ to ensure responsible innovation that aligns with societal values?

The Responsible Design framework by Centre for Humane Tech offers a comprehensive guide to approach ethical considerations in design— https://www.humanetech.com/. It provides a holistic set of principles to operate from such as "Strengthen Existing Brilliance"—which is to enhance a user's own strengths in the interactions they have with the systems, "Nurture Mindfulness"—that aims to answer the question "How can technology help increase capacities for concentration, clarity, and equanimity?" and "Bind growth with responsibility"—that forces the designer to think of what harm may the design they are building cause to the society when it scales.

And there are several inspirations to draw from—such as "Elements of Value Pyramid" from Bain & Co., or from the design principles of La Victoria Lab of Intercorp, Peru. Design gurus such as Don Norman with his Design X framework and Alan Cooper with his "ancestry thinking" framework are all there for responsible designers to learn from.

When I led the SBI digital banking project, we did explore these themes. And we got some interesting insights. For example, we tried to plot natural strengths as we explored the various user types. We redefined the problem statements and design prompts for their Insurance products (Help in tracking/changing eating habits, incentivizing healthy habits) and their investment products (balancing financial growth with leading a stress-free life). These ideas came about because we used the above frameworks. They are not yet used in mainstream design, but I hope they become mainstream soon enough.

FUTURE OF DESIGN

Considering the rapid technological advancements, what emerging trends do you foresee in design for societal impact? How should designers adapt to remain effective in addressing future social challenges?

I believe the designer of the future is a facilitator of relationships—not just between users and products, but between systems and ethics, between policy and lived experience, between the map and the territory.

Emerging tech like AI will require us to rethink speed. The challenge is not whether AI can help us analyze 500 user interviews overnight. It's whether we, as designers, can still find time to sit under a tree and listen to one user without bias.

I'm also excited about AI not as a creator, but as a collaborator.

There are challenges in this path, though. The deep, human research without the use of AI provided an opportunity to immerse oneself in the world of the user. It was not always about finding some unseen insights, but to be able to truly empathize with the user, with the people within the ecosystem. That is at the risk of being lost with the increasing use of AI.

And then there is the problem of AI hallucinations. AI tends to be more confident than the least competent among us in some scenarios. The authority with which it proclaims hallucinations as deep and profound evidence-based statements is astonishing.

So, it's imperative for the designers to watch out for the challenges above and to retain a sense of reflection, pause, and uniquely human skills, and use AI as a collaborator and a partner, not as a creator.

ADVICE FOR ASPIRING DESIGNERS

What guidance would you offer to designers aiming to create meaningful societal impact through their work, especially those at the intersection of technology and human-centered design?

To any designer starting out, here's what I'd say:

1. Don't chase perfection; chase progress.
2. Learn to sit with contradictions—between scale and empathy, speed and depth.
3. Read widely—policy papers, business books, classic literature, metaphysics, poetry, behavioral economics, and cognitive psychology.
4. Remember that being a designer is not just a job to pay your bills, but it brings with it certain responsibilities. It puts you in charge of a systemic change and creating a world that doesn't exist yet, so always be aware of that higher responsibility and purpose.

To create better futures, we must first become better listeners.

Index

Pages in *italics* refer to figures and pages in **bold** refer to tables.